On Safeguarding the Human Person in the Time of Artificial Intelligence

Magnifica Humanitas

Encyclical Letter of the Holy Father
Leo XIV

ON SAFEGUARDING THE HUMAN PERSON IN THE TIME OF ARTIFICIAL INTELLIGENCE

Magnifica Humanitas

Pauline
BOOKS & MEDIA
BOSTON

Vatican Translation

ISBN 10: 0-8198-4615-5

ISBN 13: 978-0-8198-4615-0

Published by Pauline Books & Media, 50 Saint Pauls Avenue, Boston, MA 02130-3491

Printed in the U.S.A.

www.pauline.org

Pauline Books & Media is the publishing house of the Daughters of St. Paul, an international congregation of women religious serving the Church with the communications media.

1 2 3 4 5 6 7 8 9 31 30 29 28 27 26

Contents

Introduction

1. Humanity, created by God in all its grandeur, is today facing a pivotal choice: either to construct a new Tower of Babel or to build the city in which God and humanity dwell together. Each generation inherits the task of shaping its own era, of guiding history to become a place where the dignity of every person is safeguarded, justice is promoted and fraternity is made possible. Yet every era also runs the risk of creating an inhumane and more unjust world. Whenever humanity is in danger of marring its true identity, we Christians lift our eyes to the Incarnate God, knowing that it is "only in the mystery of the Word made flesh that the mystery of humanity truly becomes clear."[1] In Jesus Christ, this humanity in its grandeur becomes the Way, the Truth and the Life, opening the path for each of us to grow toward fullness.

2. Founded on Christ, the living stone, we experience the powerful and mysterious action of the Holy Spirit, and we

1. Second Vatican Ecumenical Council, Pastoral Constitution *Gaudium et Spes*, 22: *AAS* 58 (1966), 1042.

believe that every authentic human effort to cooperate with him for the good will be blessed by our heavenly Father, in whom we place our hope. For this reason, we can diligently contribute to every initiative that builds a more just world, and we can call others to collaborate in promoting the integral development of every human being. We wish to engage in dialogue with all men and women of our time, with whom we share in the events, questions and aspirations of humanity.[2] Together with them, we seek to identify new paths for the common good and for promoting a dignified life for all. Indeed, openness to dialogue is an integral part of the Church's vocation because, constituted in Christ as "a sacrament . . . of communion with God and of the unity of the entire human race,"[3] she recognizes history as the place where the Gospel challenges and directs human experience.

3. In this spirit, Pope Leo XIII published his Encyclical *Rerum Novarum* in 1891, the 135th anniversary of which we celebrate with deep gratitude this year. With that document, my beloved predecessor gave impetus to the reflection on society, the economy and politics, which is now known as the "Social Doctrine of the Church." When some objected that the Church should not waste energy on worldly matters, but instead focus on communicating the message of eternal life,

---

2. Cf. ibid., 11: *AAS* 58 (1966), 1033–1034.

3. Second Vatican Ecumenical Council, Dogmatic Constitution *Lumen Gentium*, 1: *AAS* 57 (1965), 5.

Leo XIII responded with realism and wisdom, saying that the proclamation of the Gospel cannot overlook the concrete lives of people.[4] Many decades have passed since then, and the Magisterium, pastors, theologians and faithful have continued to reflect on social issues in the light of the Gospel. Today, the Social Doctrine of the Church is a legacy of wisdom, where we find principles for thought, criteria for discernment and judgment, and concrete guidelines for action. Founded on Sacred Scripture and Tradition, and in engagement with the sciences, it helps us clearly interpret the challenges of the present and identify appropriate ways for living out a clear Christian witness, with joy and in service to the world. It is not an inert set of concepts, but a living *corpus* of truth that safeguards and interprets humanity's vocation to a full and just life. I therefore wish to add my own voice to this living tradition, invoking the help of the Spirit of wisdom, who has dwelt in the world since its beginning (cf. Prov 8:22–31).

The *res novae* of our time

4. While Leo XIII spoke in his time of "new things" (*rerum novarum*), today we cannot limit ourselves simply to repeating his insightful teachings. Instead, we must ask God

4. Cf. Leo XIII, Encyclical Letter *Rerum Novarum* (May 15, 1891), 22: *ASS* 23 (1890–1891), 653.

for the wisdom to interpret the great trends of our time, particularly technological advances. In recent years, it has become increasingly evident how rapidly and profoundly digitalization, artificial intelligence (AI) and robotics are transforming our world. Technology should not be considered, in itself, as a force antagonistic to humanity. On the contrary, it has formed part of our history since the beginning as "a profoundly human reality, linked to the autonomy and freedom of man."[5] Over the centuries, technological development has significantly improved the living conditions of humanity. At the same time, each phase of progress has also revealed the ambiguity of tools that can cause harm when not oriented toward the good. Today, however, we find ourselves facing a new situation. The power and prevalence of emerging technologies are interwoven into the fabric of daily life, shaping decision-making processes and deeply affecting the collective imagination: "Never has humanity had such power over itself."[6] New technologies open up a horizon extending in directions that are imaginable but not yet fully predictable. This complicates the assessment of their potential impact and the long-term effects they may have on both the dignity of individuals and the common good.

---

5. Benedict XVI, Encyclical Letter *Caritas in Veritate* (June 29, 2009), 69: *AAS* 101 (2009), 702.

6. Francis, Encyclical Letter *Laudato Si'* (May 24, 2015), 104: *AAS* 107 (2015), 888.

5. It now falls to us to face the challenges of our time with clarity of thought and responsibility. It is necessary to establish adequate regulatory tools capable of upholding justice and curbing the distorting effects of technological power. Nevertheless, the issue is not limited to regulation. As Pope Francis warned, we must realistically ask ourselves who holds this power today and how they use it: "It must also be recognized that nuclear energy, biotechnology, information technology, knowledge of our own DNA, and many other abilities which we have acquired . . . have given those with the knowledge, and especially the economic resources to use them, an impressive dominance over the whole of humanity and the entire world."[7] In the past, it was largely up to the State to guide and direct innovation. Today, however, the main drivers of development are private, often transnational, parties that are endowed with resources and the capacity to intervene that surpass those of many Governments. Technological power thus takes on an unprecedented, predominantly "private" aspect, which makes it even more challenging to discern, govern and direct such power toward the common good.

6. For this reason it is necessary to begin a shared discernment process for identifying the spiritual and cultural roots of ongoing transformations. If we focus only on contingencies, we risk letting the succession of emergencies dictate the direction of our path. We are living through a rapid phase of

7. Ibid.

transition, a "change of era," in which—while some are vying for the future of new technologies and others dedicate themselves to reflecting on the matter—most people are watching and waiting, observing from afar and merely hoping for the best. For this very reason, crucial questions impose themselves on our conscience and can no longer be avoided: Where are we going? Toward what goal do we wish to orient ourselves? What direction should we choose as a people and as a human community?

Two biblical images

7. In order to answer these questions and discern how to navigate responsibly the era of AI, I would like to bring to mind two scenes from the Bible: the construction of the Tower of Babel (cf. Gen 11:1–9) and the rebuilding of the walls of Jerusalem (cf. Neh 2–6). The story of Babel appears in the Book of Genesis, at the origins of humanity, immediately after the genealogies of Noah's sons. After settling in a plain in the land of Shinar, the people decided to build a city and a tower "with its top in the heavens" (Gen 11:4). Fearing being scattered across the earth, they sought to guarantee stability and power for themselves, and above all to "make a name" for themselves. It was an impressive feat: a single language, a single technology, a single direction. However, the project concealed a profound danger. It was a project conceived without reference to God, supported by a uniformity

that eliminated diversity and that chose homogenization over communion. When a city is built on pride and the claim to self-sufficiency, communication breaks down, languages are confused and people no longer understand each other. The result is not unity, but dispersion. Babel thus reveals the limits of any effort that, however grandiose, arises from self-affirmation, sacrifices human dignity for efficiency and aspires to reach heaven without God's blessing.

8. The Book of Nehemiah, in turn, opens at a time of great vulnerability in the history of ancient Israel. After the Babylonian exile, a portion of the people returned to Jerusalem, but the city was still in ruins, the walls collapsed and the gates burned (cf. Neh 1–2). Nehemiah, a Jew in the service of the Persian King Artaxerxes, received news of the disastrous state of his ancestral city. Before taking action, he fasted, prayed and interceded for the people. He then asked the king for permission to return to Jerusalem and, upon arriving, examined the destroyed areas in silence. He did not impose solutions from above. He convened the families, assigned each of them a section of the wall to rebuild, listened to their concerns, coordinated their efforts and addressed any opposition. The narrative shows how the city is reborn, not through the initiative of one man, but through the shared responsibility of all: men, women, priests, artisans, heads of households and young people all play a part. It is an undertaking with God at the center, which rebuilds relationships before rebuilding with stones. Thus, ancient Jerusalem rediscovers a common language—not one of

uniformity, but one of communion, namely the harmony that arises when all persons assume their own role and recognize that their strength comes from the Lord.

9. In light of these two images, the Holy Spirit challenges us today regarding our relationship with technology and the ongoing digital revolution. Scientific discoveries are talents entrusted to humanity so that they may bear fruit (cf. Mt 25:14–30). Technology has the power to heal, connect, educate and protect our common home; but it can also divide, exclude and generate new forms of injustice. In the abstract, technology in and of itself is not a solution to humanity's problems, just as it is not inherently evil. In practice, however, technology is never neutral, because it takes on the characteristics of those who devise, finance, regulate and use it. Therefore, the primary choice is not between a "yes" or "no" to technology, but rather between constructing Babel or rebuilding Jerusalem; between a power that claims to dominate the heavens and a people who work together in the presence of God to rebuild the walls of fraternal coexistence.

10. We must, then, avoid the "Babel syndrome," namely the idolatry of profit that sacrifices the weak, a uniformity that neutralizes differences, and the pretense that a single language—even a digital one—can translate everything, including the mystery of the person, into data and performance. The risk of dehumanization—of building a future that excludes God and reduces the other to a means—is an ancient and ever-new temptation that today takes on a

technical guise. Instead, let us choose the "way of Nehemiah," which highlights the importance of working together to make the City of God a safe place for returning exiles. Rebuilding today means recognizing that, precisely from the plurality of voices and visions which, even though they sometimes remind us of the confusion caused by the diversity of spoken languages, a bright possibility emerges. Indeed, this is the possibility of building together, of transforming diversity into a resource and of making listening and dialogue the common ground upon which to cultivate justice and fraternity. Within this shared task, Christians discover their unique role of guiding actions toward God so that, in his light, pluralism does not dissipate into disorder, but instead, through the practice of synodality, it becomes the space in which humanity rediscovers its solid foundations and its final end. In the Book of Revelation, John sees the New Jerusalem "coming down out of heaven from God" (Rev 21:2) as a gift for all humanity. And this vision of grace is an invitation for us Christians to work together in order to foster a peaceful, just and dignified life in community within today's "cities."

Building for the common good

11. Building a city founded on the common good implies, first and foremost, building on a firm relationship with God. It means recognizing that the truth of his love calls us to life "in all its fullness" (Jn 10:10) and communion with him. Like

Saint Augustine, we too can say, "You have made us for yourself, O Lord, and our heart is restless until it rests in you."[8] Indeed, God has inscribed in our hearts a desire for happiness that embraces all the dimensions of life. The Church, in dialogue with the men and women of our time, recognizes the urgent need to safeguard and guide this aspiration toward its deepest truth.

12. Secondly, building for the common good means accepting the limits and weakness of humanity without considering them an error to be corrected. Today, the human desire for fullness of life is at risk of being misled by deceitful goals, such as the prospect of a technology that promises to free us from all weakness, and models of wellbeing that leave behind entire populations. All too often, we place our hope in unlimited "upgrades," in forms of progress that exacerbate inequalities, and in immediate solutions incapable of healing people's wounds. As a result, while some pursue the illusion of unlimited self-assertion, many are deprived of basic necessities. The Church reminds us, with a firm yet humble voice, that true fulfilment is not achieved by eliminating weakness but through harmonious growth. It is found where freedom and responsibility are intertwined with mutual care and true solidarity, and where progress is measured by the dignity of each person and the good of all peoples.

---

8. Saint Augustine, *Confessions*, I, 1, 1: *CCSL* 27, Turnhout 1981, 1.

13. Thirdly, building a world in which everyone can flourish requires shared responsibility and courage. No one can single-handedly bear the weight of the challenges the world is facing, just as no one is so weak that they cannot play their part, for "power is made perfect in weakness" (2 Cor 12:9). All are given their own section of the wall: scientists and researchers, entrepreneurs and workers, educators and legislators, civil society, popular movements and faith communities. This is the logic of subsidiarity, which values the cooperation between generations, peoples, disciplines and cultures as the best way for fostering stability, prosperity and peace. We should not be intimidated by tensions or differences because they can become creative forces when guided by shared responsibility.

14. Finally, building for the common good requires an evangelical language. We must avoid humiliating or antagonistic words, opting rather for a clarity that sheds light and a frankness that unlocks new possibilities. We cannot condone naïve enthusiasms, nor fuel unfounded fears. Instead, let us establish standards for discernment—the dignity of the human person, the universal destination of goods, the preferential option for the poor, care for our common home and peace—and let us translate these standards into practices such as responsible planning, the assessment of human and social impact, the inclusion of the most vulnerable, the promotion of digital literacy and guiding research and industry toward justice and peace.

Remaining human

15. In the recent Ordinary Jubilee Year of 2025, we walked as pilgrims of hope and were blessed with many graces. Strengthened by these gifts, we can move forward with confidence to face the arduous tasks and demanding challenges that lie ahead. In the era of artificial intelligence, when human dignity is threatened by new forms of dehumanization, ours is the pressing duty to remain profoundly human. We must lovingly safeguard the grandeur of humanity bestowed upon us and revealed in its fullness in Christ, the splendor of which no machine can ever replace. True progress always stems from a heart open to others, an intelligence willing to listen and a will that seeks what unites rather than what separates.

16. I address this heartfelt appeal to all the Catholic faithful, to all Christians and to all men and women of goodwill. Let us not be afraid to get our hands dirty on the "construction site" of our time. Like Nehemiah, let us pray, plan wisely and work perseveringly, placing God at the forefront of our actions and the human person at the center of our choices. Thus, the "rejected stones"—the poor, the sick, the migrants and the least among us—will become the cornerstone, and a solid, welcoming common home will emerge on the earth, where love and faithfulness will finally meet, and righteousness and peace will embrace (cf. Ps 85:10). This is the blessing we implore from God; and the task that stands before us is that of being builders of communion, rather than architects of

Babel. We are to be servants of the coming Kingdom, instead of lords of towers destined for ruin. With the heart of a shepherd and a father, I ask everyone to abandon the construction of yet another Tower of Babel and to join forces in building up the common good, so that humanity will never lose its beauty, and the world once again will come to recognize the human heart as the place where God desires to dwell.

Chapter One

A Dynamic Approach Faithful to the Gospel

17. In this first chapter, I intend to present synthetically how the Social Doctrine of the Church has taken shape in the recent Papal Magisterium and in the Second Vatican Council, in order to demonstrate its dynamic character. Indeed, in each era the *res novae* require that this teaching address historical questions in the light of revealed Truth. In this regard, artificial intelligence, too, should not be considered as merely yet another theme to be studied or a crisis to be managed, but rather as a development that challenges the categories of Social Doctrine from within, calling for their further development in fidelity to the Gospel.

18. This overview, however, would not be very comprehensible if, before reflecting on the contribution of individual popes and their most relevant documents, we do not first clarify some fundamental principles concerning the way in which the Church exists in history and relates to the world.

Failing to do so would expose Social Doctrine to the risk of being perceived as an undue interference in "worldly" matters or as an external code of ethics imposed from above. In reality, it stems from a Church that walks alongside humanity, recognizing the autonomy of earthly realities and the distinction between ecclesial and political communities. Indeed, it is for this very reason that she strives to serve the common good.

A Church journeying through human history

19. The Church is present in the world as a sign of unity for the entire human family. She recognizes today's questions and challenges as the current setting in which to carry out her particular vocation of listening, dialogue and service, and of being responsive to everything concerning the lives of contemporary men and women. This involvement in people's lives helps the Church understand ever more clearly that her mission has a historical scope and entails a responsibility for the way in which social relations are built. For this reason, she cannot consider herself a stranger to the forces shaping society. On the contrary, the Church actively participates in the processes by which society grows and is organized, and she offers her own contribution to the creation of a more just and fraternal society. Pope Francis emphasized this historical dimension of the Church's mission: "No one can demand that religion should be relegated to the inner sanctum of personal

life, without influence on societal and national life, without concern for the soundness of civil institutions, without a right to offer an opinion on events affecting society."[9]

20. The Church's vocation and duty to accompany humanity in the specifics of history leads her to recognize that earthly realities possess their own proper character and order. The Second Vatican Council expressed this principle with particular precision in the Pastoral Constitution *Gaudium et Spes*, whose sixtieth anniversary we remembered and celebrated with gratitude on December 7, 2025: "If by the autonomy of earthly affairs is meant that created things and societies themselves enjoy their own laws and values . . . then the demand for autonomy is perfectly in order."[10] This affirmation shows that creation bears the imprint of an original goodness that our human outlook must preserve, cultivate and bring to fulfilment. In this regard, the Church offers herself in a way that helps to interpret reality in all its depth. She supports with humble firmness the choices that promote the dignity of every person, the cohesion of communities and the good of all. The Church thus stands alongside the world without overpowering it, so that the promise

9. Francis, Apostolic Exhortation *Evangelii Gaudium* (November 24, 2013), 183: *AAS* 105 (2013), 1097.

10. Second Vatican Ecumenical Council, Pastoral Constitution *Gaudium et Spes*, 36: *AAS* 58 (1966), 1054; cf. Decree on the Apostolate of the Laity *Apostolicam Actuositatem*, 7: *AAS* 58 (1966), 843–844.

of justice and peace that the Holy Spirit continues to sustain in the heart of humanity may come to fruition in every human endeavor.

21. Recognizing that God upholds the freedom of men and women in the unfolding of history, the Second Vatican Council affirmed the distinction between the ecclesial community and the political community, emphasizing that each must operate with full autonomy. The Church's presence in the world is also expressed through her relationship with civil society and public institutions. By engaging with these entities, the Church acknowledges the value of social and political realities and honors their specific responsibilities, supporting everything that fosters the wellbeing of individuals and strengthens the fabric of society. The Church does not claim to assume the functions belonging to the State. On the contrary, she esteems those who serve the common good, and she firmly acknowledges the responsibility that civil institutions hold within society. At the same time, the mission entrusted to the Church prompts her to address the real suffering of the men and women of our time. This closeness does not stem from an intent to supplant civil institutions, much less from an implicit criticism of their work. Rather, it stems from evangelical charity, which impels the Church to draw near to the wounds of humanity whenever they surface with greater severity. When the Church intervenes, she does so following the example of the Good Samaritan, with discretion and closeness, aware that what arises from urgent necessity cannot

become the norm, nor replace the institutional responsibilities proper to the civil community.

22. Starting from this twofold acknowledgment—the autonomy of earthly realities and the distinction between ecclesiastical and political spheres of competence—allows for a clearer understanding of the direction that the Second Vatican Council set for the Church in her relationship with the world. *Gaudium et Spes* reminds us that "it is the task of the whole People of God, particularly of its pastors and theologians, to listen to and distinguish the many voices of our times and to interpret them in the light of God's word, in order that the revealed Truth may be more deeply penetrated, better understood and more suitably presented."[11] Listening to the "many voices" is no mere sociological exercise, but instead requires spiritual discernment. Guided by the Spirit, the People of God come to recognize in cultural and social transformations both the signs of the presence of Christ, who comes and guides history toward its fulfilment, and those aberrations that obscure his face. In this way, the essential core of revealed Truth is not altered, but made explicit and adopted as a living standard for guiding concrete choices, inspiring paths of personal and communal conversion, promoting structural reforms and supporting new forms of evangelical witness in public life. History is thus understood as one of the places in which the Church

11. Second Vatican Ecumenical Council, Pastoral Constitution *Gaudium et Spes*, 44: *AAS* 58 (1966), 1065.

allows herself to be taught by the Spirit about the humanizing power of the Gospel; and she learns to develop her own teaching at the service of the dignity of every person and the good of all peoples.

The wisdom of the word of God in dialogue with the human sciences

23. The Church regards all who sincerely seek "truth, goodness and beauty" as companions on the journey, and considers them as "precious allies"[12] in defending the dignity of every person and in caring for creation. Adopting the pastoral approach of the Second Vatican Council, which invites us to listen, discern and interpret the signs of the times, and enlightened by the wisdom of the word, the Church is not afraid to encounter human knowledge. Indeed, the word of God provides reliable standards for establishing paths of justice and opening ways of reconciliation and peace among peoples. When it comes to applying these standards to the complex situations of our time, the contributions of philosophy and of the human and social sciences is essential. These disciplines help us understand and analyze cultural, economic and political dynamics more deeply. Saint John Paul II recalled that the Church welcomes the contributions of the social sciences in

---

12. Francis, Apostolic Exhortation *Evangelii Gaudium* (November 24, 2013), 257; *AAS* 105 (2013), 1123.

order "to draw from them concrete insights that help her carry out her magisterial office."[13] A dialogue with such kinds of knowledge does not diminish the power of the Gospel. On the contrary, it makes it possible to identify with greater clarity what genuinely fosters the lives of individuals and communities. Following this perspective, Pope Francis emphasized that when dealing with many specific questions, the Church does not claim to offer "a definitive opinion,"[14] but recognizes the importance of listening to scientific research and of encouraging a serious and honest debate among experts while welcoming a diversity of opinions.

24. Nourished by this fruitful dialogue between the Gospel and human knowledge, the Church has progressively developed her Social Doctrine, cultivating in history a wise patrimony marked by theological and anthropological coherence rooted in the Christian understanding of the person. Precisely because this patrimony arises from faith and a corresponding vision of reality, it does not amount to a repertoire of technical solutions or an economic or political model to be set against others. Instead, it belongs to a different order,[15] namely that of the principles that guide the interpretation of

---

13. Saint John Paul II, Apostolic Letter issued "Motu Proprio" *Socialium Scientiarum* (January 1, 1994): *AAS* 86 (1994), 209.

14. Francis, Encyclical Letter *Laudato Si'* (May 24, 2015), 61: *AAS* 107 (2015), 871.

15. Cf. Saint John Paul II, Encyclical Letter *Sollicitudo Rei Socialis* (December 30, 1987), 41: *AAS* 80 (1988), 570–572.

events and sustain an evangelical understanding of historical processes and the choices these entail. Herein lies the proper function of Social Doctrine, which does not claim to supplant the responsibilities of politics or institutions, but offers itself as a foundation for collective discernment, helping to recognize and promote whatever serves the dignity of persons, the vitality of communities and the common good.

Social Doctrine as a shared discernment

25. Understanding that the truth is a gift to be shared, not a possession to be monopolized, frees the Church from the temptation of seeking forms of presence based on power. In order to rediscover the evangelical approach of a gentle proclamation of truth that is not imposed, Saint John Paul II invited us to examine honestly the times when acquiescence was given to "intolerance and even the use of violence in the service of truth."[16] In this same vein, I too have reaffirmed that the Church "does not claim to possess a monopoly on truth,"[17] because truth is not a territory to be defended, but a good to be shared. For his part, Pope Francis expressed this same perspective in his striking phrase, "time is greater than

---

16. Saint John Paul II, Apostolic Letter *Tertio Millennio Adveniente* (November 10, 1994), 35: *AAS* 87 (1995), 27.

17. Leo XIV, *Address to the Members of the "Centesimus Annus Pro Pontifice"* Foundation (May 17, 2025): *AAS* 117 (2025), 696.

space."[18] What matters most is not occupying positions of power or defending cultural strongholds, but initiating good processes and enabling them to mature. In this way, the truth of the Gospel is not imposed from above, but grows over time within the concrete interweaving of lives, communities and cultures. This is not a truth that fears diversity, but instead welcomes and guides it. It does not eliminate conflicts, but transforms them, reuniting that which history tends to scatter. This concept can also be illustrated by the image of a multifaceted polyhedron,[19] in which the one truth of the Gospel is reflected from different angles.

26. This attitude of openness to truth, which is at the same time both one and diverse, profoundly expresses the catholicity of the Church, for she embraces the entire human family yet is also immersed in the concrete situations of peoples and cultures. The Second Vatican Council reminds us that, in virtue of this very catholicity, "each part contributes its own gifts to other parts and to the entire Church."[20] In this way, the Church grows as a whole and as individual communities thanks to a mutual exchange and to shared efforts toward an ever fuller communion. It follows, then, that the People of

---

18. Francis, Apostolic Exhortation *Evangelii Gaudium* (November 24, 2013), 222: *AAS* 105 (2013), 1111.

19. Cf. ibid., 236: *AAS* 105 (2013), 1115; Francis, Encyclical Letter *Fratelli Tutti* (October 3, 2020), 215: *AAS* 112 (2020), 1045–1046.

20. Second Vatican Ecumenical Council, Dogmatic Constitution *Lumen Gentium*, 13: *AAS* 57 (1965), 17.

God are not only gathered together from many peoples, but are also intertwined through different functions, vocations, cultures and traditions, each being called to support and enrich one another. From this perspective, Saint Paul VI acknowledged that, given the great variety of historical situations, it is unrealistic to think that the Church's Social Doctrine can propose a single response that is valid in all contexts.[21] For this reason, he invited each Christian community to interpret the reality in its own country with clarity and responsibility. The fruitful tension between the universality of the Church's mission and her local roots is an intrinsic aspect of her life, for she encompasses the whole world, while addressing the specific issues of each context as the real setting in which the Gospel takes shape.

27. In light of what has been said so far, the Church's Social Doctrine can be seen more authentically. It is not a handbook of principles and norms to be applied, but a process of shared discernment. It is born from the encounter between the eternal truth of the Gospel and the questions of history. It allows itself to be challenged by the signs of the times, and draws nourishment from the contributions of science, culture and human experience. Therefore, when the dignity of our brothers and sisters is violated, when politics fails to address the tragedies of humanity, when the economy turns against the person or

---

21. Cf. Saint Paul VI, Apostolic Letter *Octogesima Adveniens* (May 14, 1971), 4: *AAS* 63 (1971), 403.

science oversteps the limits of its competence,[22] the Church—together with other Christian denominations and believers of other religions—must make her voice heard, not in order to dominate, but to promote communion. Understood in this way, Social Doctrine becomes a theology of communion in history, a history in which the Word made flesh continues to be present through dialogue, memory and prophecy.

The development of Social Doctrine from Leo XIII to the present

28. Having outlined the way in which the Church is present in history and engages in dialogue with the world, I would now like to consider the development of Social Doctrine in the Magisterium, which has responded to the major social transformations from the nineteenth century to the present day. Naturally, I cannot do justice to the full richness of this teaching, whose fundamental principles are presented in the *Compendium of the Social Doctrine of the Church* and have been further examined by recent Magisterial teaching. Nor can I systematically explore everything that has been developed in the Encyclicals of my late venerable predecessors, especially in *Laudato Si'* and *Fratelli Tutti*. Nevertheless, I will emphasize some essential points in order to show how the present

22. Cf. Francis, Apostolic Exhortation *Evangelii Gaudium* (November 24, 2013), 243: *AAS* 105 (2013), 1118.

text stands in continuity with that tradition. I would also like to stress how, within this tradition, the unchanging core of revealed truths regarding the human person and society is constantly intertwined with a renewed capacity for listening to historical situations and for responding to contemporary issues. I will now review some of the significant stages of this development, beginning with the period inaugurated by the Encyclical *Rerum Novarum*.

The first stages of the Church's Social Doctrine

29. What we now call the "Social Doctrine of the Church" is not a spontaneous product of the modern age. Instead, it is the fruit of receiving and structuring a long tradition of ecclesial reflection on life in society, rooted in Sacred Scripture, the Church Fathers and the theological and legal developments of the Middle Ages and modern era. Although the expression "Social Doctrine of the Church" was coined by Pius XII in 1950,[23] its content began to take shape as an organic *corpus* of social teaching with Leo XIII's Encyclical *Rerum Novarum*. Confronted with the "new things" of his time—the conflict between capital and labor, the question of the workforce, and economic and social transformations—Leo XIII did not limit himself merely to acknowledging the unrest, but saw these

23. Cf. Pius XII, Apostolic Exhortation *Menti Nostrae* (September 23, 1950): *AAS* 42 (1950), 657–702.

situations as an area for the Church's pastoral mission. He exposed them to rigorous discernment, illuminating their causes and possible solutions in the light of the Gospel and an integral vision of the human person created in the image of God. Saint John Paul II regarded this approach as a "lasting paradigm"[24] of Social Doctrine: an exemplary practice through which the Church, when faced with historical changes, exercises her right and duty to examine social realities, make pronouncements about them and indicate paths for finding just solutions. In this way, the perennial contents of the faith and ancient ecclesial wisdom find expression in a living doctrine that remains faithful to the Gospel while growing in response to the "new things" of every era.

30. Leo XIII's Encyclical *Rerum Novarum* constitutes a milestone in the development of the Church's social teaching. The document places the dignity of work and of workers at the forefront of its reflection; affirms the right to a fair wage for oneself and one's family; recognizes that persons have a fundamental value that takes precedence over capital and profit; defends private property along with its indispensable societal role; esteems workers' associations; and proposes forms of cooperation between the different components of society as an alternative to the mentality of class struggle. It is not surprising, then, that Pius XI defined it as the "*Magna*

---

24. Saint John Paul II, Encyclical Letter *Centesimus Annus* (May 1, 1991), 5: *AAS* 83 (1991), 799.

Carta"[25] of Christian social action. In *Rerum Novarum*, the Church's ancient wisdom regarding the human person and life in society took on a new form capable of responding to the industrial age and offering the first major systematic framework for the Social Doctrine that would be further developed in the following decades. While many of the historical conditions described by Leo XIII have changed, at least two insights remain highly relevant today: the primacy of human labor over any mindset focused solely on finance or productivity—with the consequent attention to the people and families most susceptible to exploitation—and the inseparable link between proclaiming the Gospel and pursuing a more just social order. *Rerum Novarum* thereby continues to remind us that there is no authentic evangelization that does not also affect the structures of human society.

31. Pius XI's Encyclical *Quadragesimo Anno* was published in 1931 on the fortieth anniversary of *Rerum Novarum* at the height of a major global economic crisis, marking a further step in the Church's social teaching. Rather than limiting itself to addressing the "workforce question," it broadened its focus to encompass the overall structure of the economic and political order. The Encyclical denounces the concentration of economic power in the hands of a few; criticizes both unlimited

25. Pius XI, Encyclical Letter *Quadragesimo Anno* (May 15, 1931), 39: *AAS* 23 (1931), 189; cf. Pius XII, *Radio Message on the 50th Anniversary of "Rerum Novarum"*: *AAS* 33 (1941), 198.

competition and collectivist projects that undermine the freedom and responsibility of the individual; strongly affirms the workers' right to association; and reiterates the requirement that wages be proportionate not only to performance, but also to the needs of workers and their families. Within this framework, Pius XI systematically formulated the principle of subsidiarity, which was to become one of the cornerstones of Social Doctrine. According to this principle, whatever can be carried out by individuals, families, intermediary organizations and local communities should not be carried out by higher-level authorities. Alongside these contributions, in various interventions of his Magisterium—from the Encyclicals *Non Abbiamo Bisogno* and *Mit Brennender Sorge* to *Divini Redemptoris*—Pius XI clearly recalled the societal role of private property and denounced forms of totalitarianism that demean the dignity of the person, stifle life in society, exalt the State above its just value and discriminate according to race. At least three insights of his social teaching remain particularly relevant today: the awareness that injustice concerns not only individual behavior but also economic and institutional structures; the importance of the principle of subsidiarity, which calls for the strengthening of the fabric of associations and communities while avoiding further centralization of power; and the link between the dignity of work, fair remuneration and the genuine possibility for families to lead a dignified life.

32. In the tragic context of the Second World War, and the years of reconstruction that followed, the teachings of Pius

XII made a significant contribution to the development of Social Doctrine. This is particularly true of his Christmas radio messages, in which he outlined the framework of an international order based on justice, peace and the recognition of human dignity. In these messages, the Pope proposed a dialogue with society based on an appeal to natural law understood as a set of objective principles that precede the interests of individuals and States, and which must regulate both the internal life of nations and their mutual relations. Pius XII also attributed a decisive role to professional associations, labor unions and the various intermediary organizations in the economic and social order. He recognized these organized forms of society as an essential safeguard for civil equilibrium and for protecting the common good. He affirmed the need for a sound rule of law for guarding against the abuse of power, and he recognized democracy as a means for ensuring the proper exercise of authority. At the same time, he warned against any attempt to base law on utility or force, recalling that an international order governed by the advantage of the strongest exposes weaker peoples to oppression and fundamentally undermines trust between nations. Finally, Pius XII identified profound economic imbalances between countries as one of the factors fueling conflicts.[26] Three guidelines remain particularly significant for our own

26. Cf. Pius XII, *Address to the Sacred College of Cardinals and the Roman Prelature* (December 24, 1940): *AAS* 33 (1941), 13.

times, currently marked by new forms of global power and growing inequalities: the need for law to take precedence over interests; the awareness that economic disparities are a breeding ground for tension and violence; and the necessity of a network of associations capable of mediating between the individual and the State. These guidelines continue to provide important criteria that enable Social Doctrine to interpret the dynamics of globalization and promote a more just and peaceful international order.

The years of the Second Vatican Council

33. A new phase in the Church's social teaching began with Saint John XXIII, who placed a greater emphasis on the global dimension of social issues and the language of rights. In *Mater et Magistra*, he presented the Christian faith as a light capable of uniting heaven and earth. He recalled that, while the Church's primary mission is the sanctification and proclamation of eternal goods, she does not neglect the concrete needs of people's daily lives, and is concerned with every authentic human good.[27] Based on this unified vision of humanity, John XXIII emphasized that societal life requires a balance between the initiative of citizens and groups—who are called to organize themselves and work together—and the action of the State, which must coordinate and provide

27. Cf. Saint John XXIII, Encyclical Letter *Mater et Magistra* (May 15, 1961), 2–3: *AAS* 53 (1961), 402.

support without stifling the freedom and responsibility of individuals. Hence, he drew attention to fair remuneration for work, worker participation and the growing disparities between countries. A few years later, in *Pacem in Terris*, John XXIII addressed for the first time not only the faithful, but also all people of good will, organically linking the dignity of the person to the recognition of fundamental rights and duties, and proposing a direction for society—at the international level too—based on truth, justice, love and freedom.[28] In the present day, which is marked by widespread conflict and new forms of global interdependence, the following aspects of his thought remain particularly significant: the universal perspective of his appeal; his reference to human rights as a shared framework; and his conviction that lasting peace requires institutions and relations between peoples that are inspired by the dignity of every person.

34. The Second Vatican Council marked a turning point in the Church's understanding of herself in the contemporary world. In the Pastoral Constitution *Gaudium et Spes*, the Council presented the image of a Church that is close to humanity, engaged with the world and committed to reflecting on the concrete reality of historical situations, rather than abstract concepts. The text addresses the major issues of marriage and the family, economic and societal life, the

28. Cf. Saint John XXIII, Encyclical Letter *Pacem in Terris* (April 11, 1963), 87: *AAS* 55 (1963), 301.

political community, war and peace. It insists that economic and institutional structures are just only to the extent that they serve the integral development of the person and promote the responsible participation of all.[29] The importance of this conciliar document for the Social Doctrine of the Church lies not only in having opened up horizons for thematic reflection, but also in its method of discernment that invites us to interpret historical changes guided by the Gospel and human expertise. This approach reveals that dialogue with the world is not a tactical choice for the Church, but a concrete expression of her mission because the Gospel, like leaven, is capable of transforming the structures of society from within and forging paths toward a greater humanity. The Declaration *Dignitatis Humanae* can be included in the same context. Here, the Council recognized that religious freedom is a fundamental right grounded in human dignity that must be guaranteed by law so as to prevent people from being forced to act against their conscience or impeded from seeking and professing the truth both privately and publicly.[30] This principle is highly relevant today and continues to provide Social Doctrine with decisive criteria for protecting individuals and building pluralistic and peaceful societies.

---

29. Cf. Second Vatican Ecumenical Council, Pastoral Constitution *Gaudium et Spes*, 26: *AAS* 58 (1966), 1046–1047.

30. Cf. Second Vatican Ecumenical Council, Declaration *Dignitatis Humanae*, 2: *AAS* 58 (1966), 930–931.

35. During the Pontificate of Saint Paul VI, an understanding of peace emerged that was not reduced to the mere absence of war, but took shape within the scope of integral human development. In *Populorum Progressio*, he described development as a transition from less humane to more humane living conditions. He further understood it as a process that concerns "each person and the whole person,"[31] that is every dimension of the person and all people without exception. For this reason, Paul VI could affirm that development understood in this way is in reality "the new name for peace,"[32] because it aims to eradicate the roots of injustice and conflict and create opportunities for a more dignified life for all. The establishment of the Pontifical Commission *Iustitia et Pax* should also be seen in this light as an attempt to give stable form to this insight at the ecclesial and international levels, while bearing in mind the growing gap between rich and poor countries and the need for policies that genuinely promote more humane living conditions for all.

36. In *Octogesima Adveniens*, written on the occasion of the eightieth anniversary of *Rerum Novarum*, Paul VI applied this perspective to postindustrial society, marked by urbanization, new forms of poverty and rapid cultural changes that called into question the future of individuals and communities. Paul

---

31. Saint Paul VI, Encyclical Letter *Populorum Progressio* (March 26, 1967), 14: *AAS* 59 (1967), 264.

32. Ibid., 76: *AAS* 59 (1967), 299.

VI believed that although the Gospel was proclaimed, written and lived out in a historical and cultural context very different from our own, its message was not "outdated."[33] Instead, it offers a vision of the human person, relationships, authority and the common good that is still capable of guiding economic, political and cultural choices today. In other words, the Gospel remains relevant because it provides the criteria for recognizing what humanizes or dehumanizes and what liberates or oppresses in ever-changing situations. For the Social Doctrine of the Church, Paul VI's most demanding legacy is precisely this: as long as there are people in the world who are excluded from the development befitting human dignity, the Christian community cannot be content with a theoretical proclamation of peace. Rather, beginning where people are marginalized, it must allow the Gospel to pass judgment on those economic and political structures which—as John Paul II would later remind us—can become veritable "structures of sin."[34] As a result, no person or people will be treated as expendable in the processes of development.

The recent Magisterium

37. The rich social teaching of Saint John Paul II lies at the crossroads of the crisis of the great ideological systems of the

33. Cf. Saint Paul VI, Apostolic Letter *Octogesima Adveniens* (May 14, 1971), 4–7: *AAS* 63 (1971), 404–406.

34. Saint John Paul II, Encyclical Letter *Sollicitudo Rei Socialis* (December 30, 1987), 36: *AAS* 80 (1988), 561.

twentieth century and the onset of economic globalization. His Encyclical *Laborem Exercens*, written ninety years after the publication of *Rerum Novarum*, opened up a new avenue for reflection on work. It presents fair wages as the concrete means of verifying the justness of the entire socioeconomic system because they reveal whether the worker is treated as a person or merely as a cost of production.[35] Work is not considered simply as a problem to be dealt with or a means of generating income, but a fundamental good for the person, a principle of economic activity and the key to the entire societal question. Through work, human beings bring their freedom, creativity and capacity for cooperation into play, contributing to the cultural and moral elevation of society.[36] In light of this, the various kinds of job insecurity, fragmented career paths and automation must not be evaluated solely in terms of efficiency, but in relation to the dignity of the worker, the right to sufficient remuneration and the genuine possibility of participating in society.

38. With his Encyclical *Sollicitudo Rei Socialis*, marking the twentieth anniversary of *Populorum Progressio*, John Paul II reexamined the scourge of underdevelopment. He acknowledged the failure of numerous attempts to accelerate the economic development of poor peoples and to assist them in

---

35. Cf. Saint John Paul II, Encyclical Letter *Laborem Exercens* (September 14, 1981), 19: *AAS* 73 (1981), 625–629.

36. Cf. ibid., 10: *AAS* 73 (1981), 600–602.

the process of industrialization, noting the persistent and indeed widening gap between the world's North and South.[37] He also denounced the economic, financial and commercial mechanisms that, managed by the strongest economies, structurally favor their own interests while stifling weaker economies, and he asked that they be subjected to serious ethical, not just technical, scrutiny.[38] In this context, solidarity was understood as a concrete, shared responsibility among individuals, peoples and nations—a form of social friendship or political charity oriented toward the "civilization of love" proposed by Paul VI.[39]

39. On the centenary of *Rerum Novarum*, the Encyclical *Centesimus Annus* offered a reflection on the collapse of the Soviet system and the rise of democracy and the market economy. Saint John Paul II reiterated Pius XII's message that the Church values democracy insofar as it guarantees the effective participation of citizens, enables them to elect and peacefully replace their leaders and prevents power from being monopolized by small elite groups motivated by particular or ideological interests.[40] Likewise, the Church recognizes the positive potential of the market and private initiative

---

37. Cf. Saint John Paul II, Encyclical Letter *Sollicitudo Rei Socialis* (December 30, 1987), 14: *AAS* 80 (1988), 526–528.

38. Cf. ibid., 16: *AAS* 80 (1988), 531.

39. Cf. ibid., 31–33: *AAS* 80 (1988), 555–559.

40. Cf. Saint John Paul II, Encyclical Letter *Centesimus Annus* (May 1, 1991), 46: *AAS* 83 (1991), 850–851.

only if they remain subordinate to the moral law and are guided by the principle of solidarity, without sacrificing the most vulnerable to the rationale of profit.[41] This adds a particularly relevant legacy to the Social Doctrine of the Church. The affirmation of the link between the dignity of work, solidarity among peoples, a critical assessment of democracy and the market economy continues to provide criteria for evaluating new forms of exploitation, exclusion and crises in political representation.

40. In his social Encyclical *Caritas in Veritate*, Pope Benedict XVI sought to reassess and expand the concept of development presented in *Populorum Progressio*, interpreting it in light of globalization. He noted that such development should translate into "real growth, of benefit to everyone and genuinely sustainable."[42] That is, economic progress that is truly inclusive and respectful of the limits of creation. He reaffirmed, however, that in wealthy countries new kinds of poverty were emerging as well as unprecedented forms of exclusion, while, in poorer regions, small minorities lived in consumerist affluence alongside situations of dehumanizing poverty.[43] In addition, he observed that the new global economic and financial system, marked by a vast mobility of capital and means of production, had reduced the political

---

41. Cf. ibid., 42: *AAS* 83 (1991), 844–846.

42. Benedict XVI, Encyclical Letter *Caritas in Veritate* (June 29, 2009), 21: *AAS* 101 (2009), 656.

43. Cf. ibid., 22: *AAS* 101 (2009), 657.

power of States and their ability to influence economic processes.[44] For this reason, Benedict XVI reiterated that economic activity cannot claim to solve social problems simply through the expansion of a commercial mentality, but must be ordered toward the common good, for which the political community bears its own irreplaceable responsibility.[45]

41. Benedict XVI placed charity at the center of his analysis, stating that it "is at the heart of the Church's Social Doctrine,"[46] provided that it is always united with truth. He also noted with concern that there is a tendency to dismiss moral relevance precisely within the social, legal, political and economic fields. The originality of his contribution lies in showing that development, justice, institutions and the market are not neutral realities, but spaces where charity in truth must find historical expression. This teaching is especially relevant today in light of growing inequalities, pressures in the financial markets, the environmental crisis and a lack of trust in politics. It stands as an invitation to evaluate every model of development on its ability to be inclusive and sustainable, to rebuild the relationship between economics and politics on the common good, and to acknowledge the critical and generative role of charity in public life.

42. Pope Francis' social teaching develops along the lines of *Gaudium et Spes*, which invites us to view history through

---

44. Cf. ibid., 24: *AAS* 101 (2009), 658–659.

45. Cf. ibid., 36: *AAS* 101 (2009), 671–672.

46. Ibid., 2: *AAS* 101 (2009), 642.

the lens of human hopes and vulnerabilities, and to bring them into dialogue with the Gospel. This approach emerges with particular clarity in *Evangelii Gaudium*, where he states that the Christian proclamation has an intrinsic social dimension and calls for a Church capable of listening to the cry of the poor, migrants and victims of new forms of slavery. Francis' insistence on a synodal Church, a Church that "walks together," that seeks to read the signs of the times in the light of the Gospel and allows herself to be evangelized by the poor with whom she shares history, also fits into this perspective.[47]

43. In *Laudato Si'*, Francis provided the first significant systematic treatment of the environmental crisis in a social Encyclical, demonstrating that it is not an isolated issue, but rather the ecological aspect of the contemporary socio-economic crisis. His proposal for an integral ecology combined care for our common home with the preferential option for the poor, and strongly affirmed that "the cry of the earth and the cry of the poor"[48] cannot be separated. In this light, the universal destination of goods was brought to the forefront, alongside the critique of a technocratic paradigm that seeks to reduce everything to an object to be dominated; the defense of human labor threatened by the mindset of waste;

---

47. Cf. Francis, Apostolic Exhortation *Evangelii Gaudium* (November 24, 2013), 198: *AAS* 105 (2013), 1103.

48. Francis, Encyclical Letter *Laudato Si'* (May 24, 2015), 49: *AAS* 107 (2015), 866.

and the need for intergenerational justice. Finally, he advocated for genuine dialogue between those working in the fields of politics and finance, so that neither would become self-referential.

44. Faced with the breakdown of the social fabric, a "world war being fought piecemeal," individualistic globalization and the impact of the pandemic on community ties, Francis, in *Fratelli Tutti*, sought to revive the dream of a humanity that opts for social friendship and universal fraternity. He proposed a culture of encounter, a "better politics" capable of seeking the common good, paths of reconciliation and a world that ensures "land, housing and work for all."[49] Finally, in *Dilexit Nos*, he showed that these significant social endeavors cannot be separated from a personal relationship with Christ. Turning to the word of God, he reminded us that the truest response to the love of the heart of Jesus is concrete love for our brothers and sisters, and affirmed that "there is no greater way for us to return love for love."[50]

Interpreting history in the light of faith

45. Considering this historical overview, it is clear that the Church's Social Doctrine is not the result of a project devised

---

49. Francis, Encyclical Letter *Fratelli Tutti* (October 3, 2020), 127: *AAS* 112 (2020), 1013.

50. Francis, Encyclical Letter *Dilexit Nos* (October 24, 2024), 167: *AAS* 116 (2024), 1421.

at a desk, but rather the product of a patient process in which each pontiff—together with the Second Vatican Council—made a unique contribution in light of the "new things" of each particular era. In response to the challenges of their time, each one interpreted historical changes according to the Gospel, bringing to light different aspects of a single heritage: the dignity of the person, the value of work, the universal destination of goods, solidarity and subsidiarity, care for creation and the centrality of peace and fraternity. The result is a harmonious, though not always linear, development that is marked by different emphases, progressive insights, and, at times, changes in perspective that do not break with what came before, but allow its implications to mature. If today we can speak of a *corpus* of shared principles and criteria, it is because this faith-based interpretation of history has never been interrupted, remaining ever open to the challenges posed by each generation. It is to the great principles of Social Doctrine, which direct the discernment of believers in their personal and public lives, that I now wish to turn our attention, in order to grasp more effectively their internal coherence and capacity to guide our times.

Chapter Two

Foundations and Principles of the Social Doctrine of the Church

46. The Social Doctrine of the Church is a living reality, in dialogue with history, cultures and sciences. At the same time, it enshrines a core set of unchanging truths. For this reason, it can be considered a form of wisdom that is capable of guiding the personal and societal lives of believers even today. In this second chapter, I would like to focus on some of the foundations and principles of the Church's Social Doctrine that will help us to interpret the "new things" of our time, particularly in view of the inherent dignity of the human person. In order to protect the human person in the age of artificial intelligence, I believe that today we must once again reflect on the common good, the universal destination of goods, subsidiarity, solidarity and social justice. I am convinced that a harmonious relationship between these principles requires that they be considered collectively, so that it becomes clear how they relate to and complement each other.

47. In offering these reflections, my hope is, first and foremost, to help the lay faithful and people of goodwill rediscover their duty of implementing the above-mentioned principles in their daily lives, family relationships, work and involvement in society. Thus, they will let themselves be inspired by the aim of embodying God's love in the concrete events of life. At the same time, I would like to encourage academic institutions and universities to give fresh impetus to these principles, and to apply them in a way that will be relevant and effective in addressing the digital revolution. In this way, theological and philosophical enquiry will be able to further explore and support the Church's pastoral journey, and contribute to the Magisterium's task of enlightening the consciences of the faithful and guiding their efforts to make the life of our societies more just and fraternal.

The foundations of Social Doctrine

The human person: image of the Triune God

48. The Church's Social Doctrine brings us to the very heart of our faith: the mystery of the living God, revealed in Jesus Christ, who, as a communion of Persons—Father, Son and Holy Spirit—is love itself in relationship, expressed in the mutual gift of self and in sharing with the world.[51] As the

51. Cf. Pontifical Council for Justice and Peace, *Compendium of the Social Doctrine of the Church*, Vatican City 2004, 32.

Council recalled, human persons are called to communion with God and "can fully discover their true selves only in sincere self-giving."[52] Indeed their deepest vocation is to enter into the Trinitarian dynamic of love received and shared.

49. If the mystery of God as Love is the source of Social Doctrine, we see its most concrete expression in the face of Jesus Christ, the Incarnate Word. By becoming man, the Son of God enters our history and takes on human flesh, bringing with him the love that unites him to the Father and the Holy Spirit. In him, "the mystery of humanity truly becomes clear"[53] because his humanity is completely free, open to others, capable of building healthy and beautiful relationships and committed to the total gift of self. Those who believe in him are engaged in the great work of renewal that began with the mystery of his passion, death and resurrection, and they cooperate in building up the Kingdom of God, learning to embrace all men and women as brothers and sisters, children of one Father. In this way, both the proclamation of the Gospel and Christian life, guided by the action of the Holy Spirit, tend to bring about social consequences in the world.[54]

---

52. Second Vatican Ecumenical Council, Pastoral Constitution *Gaudium et Spes*, 24: *AAS* 58 (1966), 1045.

53. Ibid., 22: *AAS* 58 (1966), 1042.

54. Cf. Pontifical Council for Justice and Peace, *Compendium of the Social Doctrine of the Church*, 38.

50. At the heart of the Christian understanding of the human person lies the great biblical affirmation that men and women are created in the image and likeness (cf. Gen 1:26–27) of the Triune God. Created for relationship, every human person is planned and willed by God to enter into communion with him, with others and with creation. Human dignity does not depend on a person's abilities, wealth or position in life, nor on the right or wrong choices made; instead, it is a gift that precedes and transcends each person, endowed by God as an expression of his unfailing love. For this reason, the human person always remains the "way for the Church"[55] and the heart of every authentic path of integral human development.[56]

The equal dignity of all human beings

51. Saint John Paul II stated that, "this heightened sense of the dignity of the human person and of his or her uniqueness, and of the respect due to the journey of conscience, certainly represents one of the positive achievements of modern culture."[57] This statement follows the line already laid out by the Second Vatican Council, which had noted a growing

55. Saint John Paul II, Encyclical Letter *Redemptor Hominis* (March 4, 1979), 14: *AAS* 71 (1979), 284.

56. Cf. Benedict XVI, Encyclical Letter *Caritas in Veritate* (June 29, 2009), 11: *AAS* 101 (2009), 647–648.

57. Saint John Paul II, Encyclical Letter *Veritatis Splendor* (August 6, 1993), 31: *AAS* 85 (1993), 1159.

recognition of the sublime dignity of all persons, their superiority over material things and their universal and inviolable rights and duties.[58] It is important to ensure that this growth in appreciation of human dignity is not obscured by the pressure of new ideologies or very powerful interests in today's world. Among these ideologies, I consider particularly insidious the one that suggests that every person must earn or justify his or her own worth, to the point of attributing greater value to those who are more efficient or effective. From this perspective, persons end up being reduced to a means of achieving results, a resource to be used and exploited, and are no longer recognized as a proper end in themselves who should never be instrumentalized. The value of persons, however, does not depend on what they achieve or produce. There are rights that apply to everyone simply by virtue of being human, and no human power can legitimately deny or arbitrarily limit them.[59]

52. When we speak of dignity, we do not always use the word in the same way. Sometimes we refer to moral dignity, namely the way in which a person directs his or her choices and actions. At other times, we think of social dignity, which refers to a person's living conditions and the concrete respect

58. Cf. Second Vatican Ecumenical Council, Pastoral Constitution *Gaudium et Spes*, 26: *AAS* 58 (1966), 1046–1047.

59. Cf. Saint John Paul II, Encyclical Letter *Centesimus Annus* (May 1, 1991), 11: *AAS* 83 (1991), 806–807.

received from society. In other cases, we refer to existential dignity, meaning the way in which a person perceives his or her own worth and the value of life. These aspects of dignity can be enhanced or diminished. In addition to these notions, there is also the more profound and important level of ontological dignity. This is the dignity that belongs to every human being simply by virtue of existing, of having been willed, created and loved by God.[60] No sin, failure, humiliation or exclusion can diminish the profound value of a human life that God has willed and called into being.[61]

53. The fundamental dignity of each person, therefore, is neither acquired nor earned, nor does it need to be justified. The recent Declaration *Dignitas Infinita* offers a summary of the Church's thinking on this subject: "Every human person possesses an infinite dignity, inalienably grounded in his or her very being, which prevails in and beyond every circumstance, state, or situation the person may ever encounter"[62] —in other words, always and without exception. The dignity of every human being can be described as infinite, as Saint John Paul II stated,[63] for two reasons: first, because the love

---

60. Cf. Dicastery for the Doctrine of the Faith, Declaration *Dignitas Infinita* (April 2, 2024), 7: *AAS* 116 (2024), 592–593.

61. Cf. ibid., 8: *AAS* 116 (2024), 593–594.

62. Ibid., 1: *AAS* 116 (2024), 589–590.

63. Cf. Saint John Paul II, *Angelus with disabled people in the Cathedral of Osnabrück* (November 16, 1980): *Insegnamenti di Giovanni Paolo II*, vol. III/2, Vatican City 1980, 1232.

of God, who calls us to friendship with him, is infinite; and second, his love is absolutely unconditional, in the sense that, even if we search endlessly, we will never find anything that can erase or deny it.

The supreme value of human rights

54. The Church gratefully acknowledges that "the movement toward the identification and proclamation of human rights is one of the most significant attempts to respond effectively to the inescapable demands of human dignity."[64] In this regard, Saint John Paul II stated that the *Universal Declaration of Human Rights*, proclaimed by the United Nations on December 10, 1948, remains one of the highest expressions of the human conscience of our time.[65] It is "a milestone on the long and difficult path of the human race."[66] For this reason, from the Christian perspective, human rights are not an external addition to the person, but an expression of intrinsic human dignity, which the international community is called to protect and promote.

---

64. Pontifical Council for Justice and Peace, *Compendium of the Social Doctrine of the Church*, 152.

65. Cf. Saint John Paul II, *Address to the 50th General Assembly of the United Nations* (October 5,1995), 2: *Insegnamenti di Giovanni Paolo II*, vol. XVIII/2, Vatican City 1998, 731.

66. Saint John Paul II, *Address to the 34th General Assembly of the United Nations* (October 2, 1979), 7: *AAS* 71 (1979), 1148.

55. Human rights are inviolable, since they are "inherent in the human person and in human dignity."[67] Consequently, they are universal and inalienable.[68] Precisely because they are grounded in the common dignity of every man and woman, they have practical consequences and legal effects, for "it would be vain to proclaim human rights if, at the same time, everything were not done to ensure the duty of respecting them, respect by all, in all places and for all."[69] Among these rights, the first is the right to life, from conception to its natural end,[70] without which it is impossible to exercise any other right. When this fundamental right is denied—as in the cases of induced abortion, killing of the innocent and euthanasia—we are faced with choices that the Church considers gravely wrong.[71]

56. Looking at our own time, we cannot ignore the fact that the protection of human rights has been exposed to two

---

67. Saint John Paul II, *Message for the 32nd World Day of Peace* (January 1, 1999), 3: *AAS* 91 (1999), 379.

68. Cf. Saint John XXIII, Encyclical Letter *Pacem in Terris* (April 11, 1963), 5: *AAS* 55 (1963), 259.

69. Saint Paul VI, *Message to the International Conference on Human Rights* (April 15, 1968): *AAS* 60 (1968), 285.

70. Cf. Saint John Paul II, Encyclical Letter *Evangelium Vitae* (March 25, 1995), 2: *AAS* 87 (1995), 402.

71. Cf. Second Vatican Ecumenical Council, Pastoral Constitution *Gaudium et Spes*, 27: *AAS* 58 (1966), 1047–1048; cf. Saint John Paul II, Encyclical Letter *Veritatis Splendor* (August 6, 1993), 80: *AAS* 85 (1993), 1197–1198; cf. Saint John Paul II, Encyclical Letter *Evangelium Vitae* (March 25, 1995), 7–28: *AAS* 87 (1995), 408–427.

particularly serious dangers. The first is that these rights are declared in a purely formal sense, while technological progress continues alongside covert or overt violations of human dignity. The second, which is in fact the root of the first, is the inability to recognize the foundation of their universality, since we have abandoned "the search for the solid foundations sustaining our decisions and our laws."[72] Pope Francis urged us not to underestimate this last issue. He pointed out that when reason seriously examines human nature, it is capable of discovering values that apply to everyone, since they derive from human nature. If this task of inquiry were abandoned, it is conceivable that rights considered untouchable today might, in the future, end up being questioned or denied by those in power, perhaps after having obtained only an apparent consensus from populations that are frightened or manipulated.[73]

57. Along with a greater awareness of the value of every human person and their rights, recognition of minority rights has also grown. Yet, there is still a long way to go to ensure that the rights of a great many, namely women, are equally and genuinely guaranteed throughout the world. It is a fact that "doubly poor are those women who endure situations of exclusion, mistreatment and violence, since they are

---

72. Francis, Encyclical Letter *Fratelli Tutti* (October 3, 2020), 208: *AAS* 112 (2020), 1043.

73. Cf. ibid., 209: *AAS* 112 (2020), 1043–1044.

frequently less able to defend their rights."[74] It is, therefore, not enough to state simply that men and women have equal dignity and rights; it is necessary that this be reflected in concrete decisions, such as in laws, access to employment, education, social and political responsibilities, and the way society listens to and values women's contributions. As long as this gap persists, we cannot say that society truly and fully recognizes that women have the same dignity as men.

58. It is individuals that matter, each and every person, together with their families. Social movements, communal ideologies and grand political proclamations in favor of a population are worthless unless they lead to the flourishing of persons—men and women—with their inalienable rights. Similarly, it is not enough to extol individual freedom or private enterprise if we then allow a multitude of people to continue living without decent work, protections or access to basic necessities.

The principles of Social Doctrine

The principle of the common good

59. Recognizing that every man and woman possesses an inalienable dignity, together with rights that no human power

74. Ibid., 23: *AAS* 112 (2020), 977. Cf. Apostolic Exhortation *Evangelii Gaudium* (November 24, 2013), 212: *AAS* 105 (2013), 1108.

can betray or nullify, requires us to shape the way we live together, including our economic and political choices, and the makeup of our cities. From this arises the first major principle of Social Doctrine that I wish to highlight: the common good. We can describe it as the social expression of the dignity recognized in every person. When Benedict XVI referred to the non-negotiable values that the Church must always defend, he included among them "the promotion of the common good."[75] For a Christian, going beyond the narrow confines of one's own interests and committing oneself, within the limits of one's ability, to the common good is a non-negotiable value, as is the promotion of life.

60. The Second Vatican Council affirmed that the common good consists in "the sum total of social conditions which allow people, either as groups or as individuals, to reach their fulfillment more fully and more easily."[76] This definition provides us with a valuable initial reference point, because the common good cannot be reduced to a mere list of conditions or institutions. It is not the sum total of individual benefits, nor the intersection of their particular interests; it is a greater good that belongs to everyone, and it can only be achieved, nurtured and protected by our collective efforts. We can say

75. Benedict XVI, Apostolic Exhortation *Sacramentum Caritatis* (February 22, 2007), 83: *AAS* 99 (2007), 169.

76. Second Vatican Ecumenical Council, Pastoral Constitution *Gaudium et Spes*, 26, *AAS* 58 (1966), 1046–1047.

that social action reaches its fullness when it is directed toward this shared good, just as a person's moral action finds its fulfillment in the choice of the true good.[77]

61. In this sense, we can say that the whole is "greater than the sum of its parts"[78] and that, for this very reason, "the mere sum of individual interests is not capable of generating a better world for the whole human family."[79] Indeed, it is an illusion to think that simply pursuing one's own progress without caring for others is sufficient for contributing to the good of all. This view ignores the inherent and specific value of the common good, which is the result of an "interdependence"[80] that creates a network of social good that expands and has an impact on people. The common good is a "*plus*," the result of interaction and mutual influence that connects various actions, initiatives, efforts and decisions. If we were to add up the individual goods, we could not explain the existence of this "*plus*" that transcends them and, at the same time, enriches them.

62. It is the pursuit of the common good that gives life to a people, understood not as a mere collection of individuals,

---

77. Cf. Pontifical Council for Justice and Peace, *Compendium of the Social Doctrine of the Church*, 164.

78. Francis, Apostolic Exhortation *Evangelii Gaudium* (November 24, 2013), 235: *AAS* 105 (2013), 1115.

79. Francis, Encyclical Letter *Fratelli Tutti* (October 3, 2020), 105: *AAS* 112 (2020), 1005.

80. Saint John Paul II, Encyclical Letter *Sollicitudo Rei Socialis* (December 30, 1987), 38: *AAS* 80 (1988), 564.

but as a living reality in which people learn to recognize that they themselves are interconnected and jointly responsible for the *res publica*. In this sense, every person contributes to the building up of one's people through "a slow and arduous effort calling for a desire for integration and a willingness to achieve this through the growth of a peaceful and multifaceted culture of encounter."[81] Working together for the common good means having a shared vision. It is clear that there are many ideological and practical differences among people, as well as differing interests and frequent disagreements, but that does not mean it is impossible to engage in dialogue to establish a set of basic agreements that enable the creation of a shared vision, upon which everyone can move forward together.

63. It is the State's responsibility to ensure cohesion, unity and the proper organization of civil society, so that the common good can be pursued with everyone's contribution. In practical terms, this means that public authorities have the delicate duty to "harmonize the different sectoral interests with the requirements of justice,"[82] seeking a balance between individual interests and the common good, without leaving behind the most vulnerable. When politics abandons a long-term perspective and reduces itself to short-term calculations or sterile polarizations, then the language of the

81. Francis, Apostolic Exhortation *Evangelii Gaudium* (November 24, 2013), 220: *AAS* 105 (2013), 1110.

82. Pontifical Council for Justice and Peace, *Compendium of the Social Doctrine of the Church*, 169.

common good loses credibility, and, at the same time, social inequalities and divisions grow.

64. This also applies to international politics. As the divide between nations widens, a mentality of confrontation and aggression begins to take hold, and the difficult path toward a more united and fraternal world suffers new and painful setbacks. In this context, speaking of a shared journey toward a more just development for the entire human family "sounds like madness."[83] Yet we must not lose hope. I invite everyone to conceive of ways of cooperating and of more effective international institutions, capable of safeguarding the global common good without compromising the legitimate diversity of peoples and nations. Indeed, the promotion of the common good can never be separated from respect for the right of peoples to exist, to preserve their own identity and to contribute their unique qualities to the family of nations.[84] Moreover, any attempt or plan to eliminate or subjugate a nation is gravely immoral and therefore unacceptable.

The principle of the universal destination of goods

65. "Among the numerous implications of the common good, immediate significance is taken on by the principle of

---

83. Francis, Encyclical Letter *Fratelli Tutti* (October 3, 2020), 16: *AAS* 112 (2020), 974.

84. Cf. Saint John Paul II, *Address to the 50th General Assembly of the United Nations* (October 5, 1995), 8: *Insegnamenti di Giovanni Paolo II*, vol. XVIII/2, 735.

the universal destination of goods."[85] First of all, this principle reminds us that the earth's goods—soil, water, air and natural resources—are given by God to the entire human family to sustain the lives of all, and that every person has an inherent right to the use of such goods, both now and in the future. Saint John Paul II recalled that, "God gave the earth to the whole human race for the sustenance of all its members, without excluding or favoring anyone."[86] Consequently, "it is not in accordance with God's plan to use this gift in such a way that its benefits accrue solely to a select few."[87] Today, we are called to recognize that this universal destination applies not only to material goods, but also to immaterial and cultural goods.

66. Certainly there is a right to private property, which has its own specific meaning and purpose, yet it is always subordinate to the universal destination of goods. According to John Paul II, this subordination is the golden rule of social conduct and the "first principle of the whole ethical and social order."[88] In the Church's tradition, property has

---

85. Pontifical Council for Justice and Peace, *Compendium of the Social Doctrine of the Church*, 171.

86. Saint John Paul II, Encyclical Letter *Centesimus Annus* (May 1, 1991), 31: *AAS* 83 (1991), 831.

87. Saint John Paul II, *Homily during the Mass celebrated for farmers at Recife* (July 7, 1980), 4: *AAS* 72 (1980), 926.

88. Saint John Paul II, Encyclical Letter *Laborem Exercens* (September 14, 1981), 19: *AAS* 73 (1981), 626.

been viewed as a means of protecting and managing goods so that they may better serve the common good. Since "the Christian tradition has never recognized the right to private property as absolute or inviolable,"[89] its social function must not be considered a mere theological opinion, but a doctrine of the Church, already present in Sacred Scripture and in the writings of the Church Fathers. For this reason, Pope Francis reminded us that solidarity, when lived out in its fullest sense, also means "to restore to the poor what belongs to them."[90]

67. Today, among the goods that are universally intended for everyone, we must also include new forms of property, such as patents, algorithms, digital platforms, technological infrastructure and data. In a context where the wealth of nations depends increasingly on knowledge and technology, when these goods remain concentrated in the hands of a few, without adequate forms of sharing and access, a new imbalance is created that contradicts the universal destination of goods. In turn, it widens the gap between the included and the excluded, between those who can participate in the digital revolution and those who remain on the margins. Furthermore, care for our common home and our responsibility toward the

---

89. Francis, Encyclical Letter *Laudato Si'* (May 24, 2015), 93: *AAS* 107 (2015), 884; cf. Encyclical Letter *Fratelli Tutti* (October 3, 2020), 120: *AAS* 112 (2020), 1010.

90. Francis, Apostolic Exhortation *Evangelii Gaudium* (November 24, 2013), 189: *AAS* 105 (2013), 1099.

poor and future generations require that the use of the goods of creation and the new possibilities offered by technology be regulated in such a way as to respect the environment, avoid waste and prevent new forms of exploitation.

The principle of subsidiarity

68. The principle of subsidiarity stems from the very same understanding of the human person that has guided our reflection on dignity and the common good. If every woman and man is called to take ownership of his or her own life and to contribute to the formation of society, then social institutions must also respect and support this responsibility. The Social Doctrine of the Church refers to subsidiarity as the principle according to which the role of individuals, families, local communities and intermediary organizations should not be supplanted by higher-level authorities. Moreover, higher-level institutions must recognize, protect and promote the freedom and creativity of lower-level entities, coordinating their contributions so that they can cooperate effectively for the common good.[91]

69. Starting with Leo XIII and the beginnings of modern social teaching, the Church has insisted that neither the individual nor the family should be subsumed by the State, but

---

91. Cf. Pontifical Council for Justice and Peace, *Compendium of the Social Doctrine of the Church*, 187.

should be allowed to act freely, as far as possible, without harming the common good.[92] Saint John Paul II took up and developed this perspective, noting that the political community is at the service of civil society and that the State must protect the common good, intervening when necessary, but without permanently supplanting the responsibilities of intermediary organizations and social institutions.[93] Subsidiarity does not justify the State's disengagement, but rather guides its actions. Indeed, public intervention is necessary precisely to enable all social actors to fulfill their mission without being stifled. It is the responsibility of the political community to create the conditions that allow individuals, families, associations and intermediary organizations to fulfil their mission in society, without being replaced or reduced to mere facilitators.[94]

70. This principle encourages us to move beyond any form of paternalistic or welfare-based management of societal life, but instead to promote a culture of shared responsibility in a State that values citizens' initiative, and a civil society capable of forging bonds and mobilizing energies in the service of the common good. In accordance with the principle of subsidiarity, decisions are made at the closest level possible to the

92. Cf. Leo XIII, Encyclical Letter *Rerum Novarum* (May 15, 1891), 26: *ASS* 23 (1890–1891), 656.

93. Cf. Saint John Paul II, Encyclical Letter *Centesimus Annus* (May 1, 1991), 11: *AAS* 83 (1991), 806–807.

94. Cf. ibid.

persons involved, thereby fostering community life and avoiding people being presented with decisions that have already been taken. In this way people can participate in the decision-making process. When families, associations, local communities, volunteer organizations and those in the so-called "third sector" are recognized and supported, social life becomes more accessible to people, services become more attuned to real needs, and solutions are more creative and respectful of the dignity of each person.[95]

71. The principle of subsidiarity applies especially in the context of the digital revolution. Here, the highest level is not the State, but rather major economic and technological actors that exercise *de facto* power over the conditions of everyday life. This level, which monopolizes expertise, data and decision-making authority, involves companies and platforms that define conditions for access, rules of visibility, forms of interaction, and even economic opportunities. The principle of subsidiarity requires that such processes not be imposed from above in an opaque and unilateral manner, but instead be directed toward the common good with transparency, accountability and meaningful forms of participation (including independent checks, transparency regarding algorithms, equitable access to data and avenues for recourse).[96]

---

95. Cf. ibid., 48: *AAS* 83 (1991), 852–854.

96. Cf. Francis, Encyclical Letter *Fratelli Tutti* (October 3, 2020), 169: *AAS* 112 (2020), 1028.

72. In this context, States and transnational institutions are called to ensure fair rules and effective safeguards, so that local communities, intermediary organizations, schools, universities, religious institutions and associations have a voice and can contribute to the discernment of choices that affect people's daily lives, such as employment, access to services, data management and digital environments. When it comes to decisions regarding economic flows and digital platforms, as well as the governance of data and algorithms, we cannot allow a handful of actors to dictate these processes on their own; instead, we must build forms of cooperation that respect the various levels of the global community and make them jointly responsible for the common good.[97]

The principle of solidarity

73. Having considered the common good and subsidiarity, I would like to reflect on the principle of solidarity. This emerges from a vision of the human person generated by faith, namely that every human being is created in the image of God and is part of a network of relationships that bind him or her to others, to specific populations and to creation. Saint Paul VI observed that the obligations of solidarity, justice and charity are rooted in the human and supernatural

---

97. Cf. ibid., 168: *AAS* 112 (2020), 1027–1028.

fraternal bonds that unite individuals and populations.[98] Fraternity is not merely an aspiration of believers, but is a social and political reality to be embodied in communal choices and endeavors. Solidarity, then, is the concrete recognition that the future of each individual is connected to the future of all; indeed, "no one is saved alone."[99] The close link between subsidiarity and solidarity thereby becomes evident. It is thus clear that there is an intimate link between subsidiarity and solidarity. When subsidiarity is not linked to solidarity, it ends up becoming merely the protection of particular interests; when solidarity is not supported by subsidiarity, it degenerates into a form of welfare that does not foster responsibility.[100] This interconnectedness also pertains to the responsibility of authentic participation. Solidarity is expressed when each person, both individually and collectively, takes part in the life of the community—by staying informed, engaging with others, making their voice heard and contributing to public decisions and choices—while also assuming real responsibility so that the common good is achieved through shared decision-making.

---

98. Cf. Saint Paul VI, Encyclical Letter *Populorum Progressio* (March 26, 1967), 17: *AAS* 59 (1967), 265–266.

99. Francis, Encyclical Letter *Fratelli Tutti* (October 3, 2020), 32 and 54: *AAS* 112 (2020), 980 and 988.

100. Cf. Benedict XVI, Encyclical Letter *Caritas in Veritate* (June 29, 2009), 58: *AAS* 101 (2009), 693–694.

74. In many areas, we are already experiencing a kind of "*de facto* solidarity," for our lives are intertwined; digital networks connect people and communities across the world in real time, and global economies and communications mean that events in one place have a far-reaching impact. This network of relationships, however, only constitutes solidarity in the fullest sense of the word when it becomes a conscious choice. Faith invites us to see this reality as a call: we are not merely neighbors to one another, but entrusted to each other, so that each of us may take responsibility, as best we can, for the lives and wounds of our brothers and sisters. Solidarity arises precisely when we decide not to remain indifferent to what happens to our neighbor but instead to transform unavoidable bonds—economic, cultural and technological—into paths of sharing, cooperation and mutual care, embracing the idea of "thinking and acting in terms of community."[101]

75. The Church's social teaching emphasizes that solidarity is both a principle and a virtue. As a principle, it expresses the objective order of relationships among individuals, groups and peoples, pointing to an awareness of interdependence whereby the good of each person depends on the good of others. As a virtue, it requires a "firm and persevering determination"[102] to strive for the common good, with particular

---

101. Francis, Encyclical Letter *Fratelli Tutti* (October 3, 2020), 116: *AAS* 112 (2020), 1009.

102. Saint John Paul II, Encyclical Letter *Sollicitudo Rei Socialis* (December 30, 1987), 38: *AAS* 80 (1988), 564.

attention to those most in need. Pope Francis noted that solidarity is "a way of making history"[103] that creates communities and not just masses of individuals. For this reason, it requires a modest and shared way of life, the ability to forego immediate benefits in order to create opportunities for others in the future, and a willingness to challenge habits and privileges—including those related to digital consumption and the use of technology—when they prevent others from living with dignity.

76. In a world marked by increasingly close connections between people, communities and nations, solidarity also takes on a global dimension. Benedict XVI strongly emphasized the link between development, justice and responsibility toward future generations, stating that authentic development requires solidarity and inter-generational justice,[104] as well as an awareness of the bonds that unite us to the natural environment. Today, this responsibility also extends to digital and information infrastructure. Like the natural environment, the "digital ecosystem" can be preserved or exploited, shared or monopolized. Solidarity demands that decisions regarding data, algorithms, platforms and artificial intelligence take into account not only the immediate benefit for a few, but also the impact on all peoples and on future generations.

---

103. Francis, Encyclical Letter *Fratelli Tutti* (October 3, 2020), 116: *AAS* 112 (2020), 1009.

104. Cf. Benedict XVI, Encyclical Letter *Caritas in Veritate* (June 29, 2009), 48: *AAS* 101 (2009), 685.

The principle of social justice

77. For the Christian community, social justice is a concrete way of following Jesus and remaining faithful to the Gospel. In the New Testament, Jesus proclaims the "good news to the poor" (Lk 4:18) and identifies himself with the lowly, the sick, the imprisoned and strangers (cf. Mt 25:31–46). He thus teaches us that justice is born from, and fulfilled in, fraternity, because the way we approach and relate to the least among us becomes, in concrete terms, the measure of our relationship with God and with our brothers and sisters. Justice, however, concerns not only the behavior of individuals, but also the way in which the structures of society are conceived and organized. In this regard, the Second Vatican Council reminds us that every institution is called to serve the human person and his or her dignity.[105] Social justice is, therefore, characterized by the capacity of a social, economic and political order to allow everyone—particularly the weakest—to live a truly dignified life, without leaving anyone behind.

78. The recent Magisterium has insisted that social justice begins with the least among us. Saint John Paul II spoke of a preferential option for the poor[106] that must guide both

105. Cf. Second Vatican Ecumenical Council, Pastoral Constitution *Gaudium et Spes*, 25: *AAS* 58 (1966), 1045–1046.

106. Cf. Saint John Paul II, Encyclical Letter *Sollicitudo Rei Socialis* (December 30, 1987), 42: *AAS* 80 (1988), 572–574.

personal and societal choices, while Pope Francis denounced a "'throw away' culture"[107] that generates ever new forms of exclusion. From this perspective, social justice requires us to look at individuals and communities, starting with the most vulnerable: the poor, migrants, refugees, internally displaced persons, victims of violence and people living in urban or existential peripheries.

79. The idea of "social justice" helps us recognize that injustices do not arise solely from the wrong choices of individuals, but also from structures, mechanisms and economic and cultural systems that produce inequality almost automatically. Saint John Paul II spoke in this vein of structures of sin[108] that oppose God's will and require a commitment to personal and social conversion. In this perspective, justice is not merely about the fairer distribution of resources or the correction of current injustices, but also assumes a restorative dimension. It aims to mend broken bonds and reintegrate those who have been excluded, taking into account the wounds caused by injustices, such as wars, colonialism, racial or gender discrimination, violence against entire peoples and exploitation. This may include restoring dignity and a voice to those who have been ignored, fostering processes of

---

107. Francis, Apostolic Exhortation *Evangelii Gaudium* (November 24, 2013), 53: *AAS* 105 (2013), 1042.

108. Cf. Saint John Paul II, Encyclical Letter *Sollicitudo Rei Socialis* (December 30, 1987), 36–37: *AAS* 80 (1988), 561–564.

healing for collective memory, opposing discriminatory laws and practices, and providing concrete support to those who still bear the consequences of wrongs suffered in the past.

80. In this day and age, social justice must also grapple with the environment shaped by digital technologies. The spread of global networks, platforms and artificial intelligence systems is changing the way we obtain information, communicate and access services. Justice demands that we prevent the emergence of new forms of exclusion and deprivation of freedoms: individuals and peoples hindered or denied access to basic technologies, communities exposed to invasive surveillance and social groups penalized by opaque algorithms that perpetuate prejudice and discrimination. In the digital age, a just social order guarantees everyone equal access to opportunities, protects the youngest and weakest members of society, combats hate and misinformation and subjects the use of data and technology to public oversight, so that the guiding principle is not solely profit but the dignity of every person and the common good of all people.

81. A litmus test for social justice today is the treatment of migrants, refugees and those forced to move due to poverty, violence, climate change and environmental disasters. The way a society treats them reveals whether its sense of justice is driven by fear or by the spirit of fraternity. Pope Francis urged us to see migrants not simply as a problem to be managed, but as a living image of the People of God on the

move.[109] They are people with dignity, resources and dreams, who have the right to be treated with respect and to ask to become active members of the societies that welcome them. Social justice in this area entails at least two complementary commitments. On the one hand, this means protecting the rightful hopes of those forced to leave by ensuring safe and legal routes, dignified conditions for receiving them, and genuine pathways to integration. On the other hand, it means promoting the right to remain in one's homeland in peace and security by addressing the root causes that force people to migrate, including those linked to economic injustices and the climate crisis. When these rights are respected, migration can become an opportunity for encounter and mutual enrichment among peoples.

Integral human development

82. In his Encyclical *Populorum Progressio*, Paul VI affirmed that development is authentic only if it is "integral," meaning that it can "foster the development of each man and of the whole man."[110] In the decades that followed, the Social

---

109. Cf. Francis, *Message for the 110th World Day of Migrants and Refugees* (September 29, 2024): *AAS* 116 (2024), 735.

110. Saint Paul VI, Encyclical Letter *Populorum Progressio* (March 26, 1967), 14: *AAS* 59 (1967), 264.

Doctrine of the Church reprised and reflected on this expression in order to indicate the practical ways in which the noble principles—dignity, the common good, the universal destination of goods, subsidiarity, solidarity and social justice—are implemented in real life. By "integral human development," we mean a process in which the growth of individuals and peoples encompasses all dimensions of existence and opens the future to subsequent generations as well.

83. For individuals as well as for nations, development is both a duty and a right. Minimum conditions are required for enabling every person and people to flourish in accord with their dignity, without being kept in a state of dependence or excluded from access to necessary goods. Development is truly human when it places people at the center instead of the accumulation of wealth, and when it concerns peoples as well as individuals. Justice demands the recognition of the rights of society and the rights of peoples, and includes a responsibility toward future generations. Development is not truly human if it increases consumption for some while shifting costs and burdens onto others, or relegates entire regions to subordinate roles, preventing them from realizing their full potential.[111] Development is integral when it is not limited to the economic sphere, but promotes quality of life in its spiritual, cultural, moral and relational dimensions, while

111. Cf. ibid., 17: *AAS* 59 (1967), 265–266; Francis, Encyclical Letter *Fratelli Tutti* (October 3, 2020), 125–127: *AAS* 112 (2020), 1012–1013.

respecting our common home, the diversity of peoples and their ways of life.[112]

84. Today, the concept of integral human development is a benchmark for the evaluation of integral ecology, which has become an indispensable dimension of the Church's Social Doctrine. Indeed, the quality of development is measured by the ability to integrate justice toward people and the care of our common home, and to promote dignified living conditions, access to necessary goods, just social relations, care of creation and consideration for future generations. It follows that true progress is not what increases the wellbeing of some by degrading ecosystems, shifting costs onto the most disadvantaged communities, or compromising the living conditions of those who will follow us.

85. Seen in this light, integral human development is the framework through which we can interpret the changes of our time, including those brought about by the digital revolution. Technological innovations, including artificial intelligence, are not neutral, for they can either foster participation and justice or exacerbate inequality, control and exclusion. For this reason, they must be evaluated by asking a crucial question: Do they

---

112. Cf. Saint Paul VI, Encyclical Letter *Populorum Progressio* (March 26, 1967), 14: *AAS* 59 (1967), 264; Benedict XVI, *Address to the Diplomatic Corps accredited to the Holy See* (January 8, 2007): *AAS* 99 (2007), 73; Francis, *Address to Participants of the 3rd Global Meeting of the Indigenous Peoples' Forum of the International Fund for Agricultural Development* (February 15, 2017): *AAS* 109 (2017), 244–245.

truly help individuals and peoples to become more humane and fraternal, while respecting our common home and future generations? It is here that the principles of Social Doctrine become concrete criteria for discernment regarding the issues which we will address in the following chapters.

An examen for the Church

86. In conclusion, I would like to touch on a point that is particularly close to my heart. Social Doctrine is not merely a message addressed to society; it is also an examination of conscience for the Church—a home and school of communion that is always called to ensure that the principles outlined in this chapter are applied, especially within its own structures. In the ecclesial context, the common good takes the form of a synodal approach for mission at the service of the Kingdom. Indeed, the Church is the "communitarian and historical subject of synodality and mission."[113] This requires attention to the way decisions are taken and responsibilities are exercised. The *Final Document* of the Synod identifies a culture of transparency, accountability and evaluation as key practices for missionary transformation.[114]

87. With this in mind, subsidiarity becomes the guiding principle for governance and pastoral life. It involves

113. *Final Document of the Second Session of the XVI Ordinary General Assembly of the Synod of Bishops* (October 26, 2024), 17.

114. Cf. ibid., 11.

recognizing and supporting the faithful and intermediary ecclesial organizations as they carry out their responsibilities, valuing charisms and skills and avoiding any form of paternalism that suffocates evangelical freedom. In practical terms, the participation of the baptized in decision-making processes and their shared responsibility in the mission are achieved through genuine, rather than merely nominal, participatory bodies.[115]

88. For the Christian community, solidarity finds its source in the mystery of Christ and is nourished by the Eucharist. Solidarity emerges from communion in faith and the Sacraments: Baptism and Confirmation unite us in Christ, so that we may become one Body and one Spirit, one heart and one soul (cf. Eph 4:4; Acts 4:32). The Eucharist, which is the sacrament of unity, nurtures our belonging to the Body of Christ and teaches us how to share. The diverse sensibilities present in the Church and the strong convictions that animate each person are a source of richness if they remain anchored in the certainty that unity is a gift received and a responsibility to be fulfilled.

89. Living out justice in the Church means purifying ecclesial relationships and structures from distortions that give rise to inequality, lack of transparency and abuse of power. In this regard, listening to the victims of spiritual, economic, institutional, sexual and power-based abuse, as well

115. Cf. ibid., 103–108.

as abuses of conscience, is an integral part of a journey toward justice, which includes acknowledging the harm done, just reparation and taking steps to prevent it from happening again. Every power is at the service of communion and mission. All authority is at the service of the People of God. This ministry of service is expressed not only through our faith celebrated and lived in the Sacraments, and in the adoption of a synodal style, but also in the concrete sharing of goods. Following the example of the early Church, ecclesial resources need to be shared so that no one among us may be in need (cf. Acts 4:34), and so that their administration may support the mission of proclaiming the Gospel to the poorest. Regular assessments of the exercise of ministerial responsibilities should be encouraged, not as judgments on individuals, but as tools for learning and correction oriented toward mission.[116] Only to the extent that we are open to the action of the Holy Spirit will these principles of Social Doctrine become incarnate in ecclesial life. In this way, the Church will be able to bear credible witness to society that seeking the common good together, with shared responsibility and fraternity, is not a utopia, but a real possibility.[117]

---

116. Cf. ibid., 100–101.

117. Cf. Francis, Encyclical Letter *Fratelli Tutti* (October 3, 2020), 94: *AAS* 112 (2020), 1001.

Chapter Three

Technology and Dominance

The Grandeur of Humanity in Light of the Promises of AI

90. Having recalled the principles that shine a light on Social Doctrine, I would now like to focus on certain challenges that profoundly shape our way of living today. The biblical image accompanying these reflections is that of a building project. On the one hand, there is the Tower of Babel, where collective effort follows a plan that dominates and ultimately dehumanizes (cf. Gen 11:1–9). On the other hand, there are the ruins of Jerusalem, which under Nehemiah's direction are rebuilt piece by piece as a project of shared responsibility (cf. Neh 2–6). We are called to reflect on the great "construction sites" of our era and ask: What are we building? As technological development rapidly transforms languages, relationships, institutions and forms of power, we believers must and can choose which projects to work on and

in what manner, so as to safeguard and value the grandeur of humanity that has been given to us as a gift. This is a choice not only for our future but also for our present, since artificial intelligence and other emerging technologies are already part of our daily lives.

91. I am convinced that the concrete way of living out social relationships in the light of the Gospel is not established once and for all, but remains a task entrusted, from generation to generation, to the Christian community. Under the guidance of the Holy Spirit, the Church allows herself to be enlightened by God's word, reads the signs of the times and creatively seeks new ways for relationships between peoples and nations to become ever more conformed to the demands of the Kingdom of God.[118] For this reason, I encourage all members of the Church not to be afraid of the present challenges, but to listen to one another and firmly embrace their responsibilities in building a more humane and fraternal society.

The technocratic paradigm and digital power

92. In his Encyclical *Laudato Si'*, Pope Francis denounced the growing dominance of a technocratic paradigm[119] in our

118. Cf. Pontifical Council for Justice and Peace, *Compendium of the Social Doctrine of the Church*, 53.

119. Cf. Francis, Encyclical Letter *Laudato Si'* (May 24, 2015), 106–109: *AAS* 107 (2015), 889–891.

globalized world: the tendency to let the logic of efficiency, control and profit alone shape personal, social and economic decisions. This makes it clear that technology is not simply a tool. When it becomes the standard by which everything is judged, it begins to dictate what matters and what can be discarded, reducing creation to an object of exploitation and human beings to mere cogs in a system driven toward ever greater efficiency.

93. This paradigm has spread rapidly in recent years, fueled in part by the expansion of artificial intelligence, cognitive science, nanotechnology, robotics and biotechnology. In themselves, these innovations can greatly serve integral human development and the care of our common home. Yet precisely because of their power, they can also hasten the expansion of the technocratic paradigm and therefore require a new spiritual, ethical and political framework. More power does not necessarily imply something better. In this respect, the words of Romano Guardini remain relevant: "Contemporary man has not been trained to use power well."[120]

94. The danger of humanity becoming a victim of its own achievements was already clearly recognized by Saint Paul VI, who warned that "the most extraordinary scientific progress, the most astounding technical feats and the most amazing economic growth, unless accompanied by authentic moral and social progress, will in the long run go against

120. R. Guardini, *Das Ende der Neuzeit*, Würzburg 1951, 89.

man."[121] For this reason, technological progress—valuable in itself—requires careful discernment of the anthropological vision that guides it and the ends it pursues. If technological development advances without a corresponding ethical and social progress, the result may be an increase in means without a growth in humanity: "having more" without "being more." In such a scenario, there is a risk that individuals will be evaluated principally according to the outcomes they produce.[122]

95. Here, we must recognize another crucial aspect, which I have noted earlier. In many cases within the digital context, control over platforms, infrastructure, data and computing power does not rest with States, but with major economic and technological actors. These entities effectively set the conditions for access, determine the rules of visibility and shape the very possibilities for participation. When such power is concentrated in the hands of a few, it tends to become opaque and evade public oversight, increasing the risk of distorted forms of development that give rise to new dependencies, exclusions, manipulations and inequalities.

96. Faced with this concentration of power in the digital world, the criteria for judgment and discernment in this new situation are the noble principles of Social Doctrine: the

---

121. Saint Paul VI, *Address on the occasion of the 25th Anniversary of the FAO* (November 16, 1970): *AAS* 62 (1970), 833.

122. Cf. Francis, *Address to the Council for an Inclusive Capitalism* (November 11, 2019): *L'Osservatore Romano*, November 11–12, 2019, 8.

inalienable dignity of the human person, the common good, the universal destination of goods, subsidiarity, solidarity and social justice. They demand that we assess whether the power of digital infrastructures and algorithms truly fosters participation and responsibility, protects the vulnerable, ensures fair access to opportunities and remains directed toward the good of all. On this basis, we can now examine more closely what artificial intelligence is, the possibilities it opens up and the risks it entails.

Artificial intelligence

97. It is not my intention here to offer a comprehensive treatment of artificial intelligence, nor to give an overview of the extensive relevant literature, since authoritative contributions already exist, including within the ecclesial context.[123] I limit myself to recalling a few essential elements for a moral

---

123. Cf. Dicastery for the Doctrine of the Faith and Dicastery for Culture and Education, Note *Antiqua et Nova* (January 14, 2025): *AAS* 117 (2025), 159–210; Francis, *Message for the 57th World Day of Peace* (December 8, 2023): *AAS* 116 (2024), 54–64; Francis, *Message for the 58th World Day of Social Communications* (January 24, 2024): *AAS* 116 (2024), 261–266; Francis, *Address to the G7 Session on Artificial Intelligence*: "An exciting and fearsome tool" (June 14, 2024): *AAS* 116 (2024), 866–875; International Theological Commission, *Quo vadis, humanitas? Thinking about Christian anthropology in the face of some scenarios on the future of humanity* (February 9, 2026); Leo XIV, *Message for the 60th World Day of Social Communications* (January 24, 2026): *L'Osservatore Romano*, January 24, 2026, 2–3.

and social discernment that safeguards the primacy of the human person, in order to ensure that it will always be human intelligence, with its conscience and freedom, that guides technical innovations and responsibly determines their use and limits.

98. It is appropriate to preface this discussion with two considerations. First, any statement regarding AI risks becoming quickly outdated, given the remarkable pace at which these systems are developing. Second, all of us, including those who design them, possess only a limited understanding of their actual functioning. Indeed, current AI systems are more "cultivated" than "built," for developers do not directly design every detail, but instead create a framework within which the intelligence "grows." As a result, fundamental scientific aspects—such as the internal representations and computational processes of these systems—remain, at present, unknown. There thus emerges an urgent need for a twofold commitment: on the one hand, a deepening of scientific research; on the other, the exercise of moral and spiritual discernment.

99. It is not possible to provide a single, comprehensive definition of AI. What can be stated, however, is that we must avoid the misconception of equating this type of "intelligence" with that of human beings. These systems merely imitate certain functions of human intelligence. In doing so, they often surpass human intelligence in speed and computational capacity, offering tangible benefits across many fields. Yet this power remains entirely tied to data processing.

So-called artificial intelligences do not undergo experiences, do not possess a body, do not feel joy or pain, do not mature through relationships and do not know from within what love, work, friendship or responsibility mean. Nor do they have a moral conscience, since they do not judge good and evil, grasp the ultimate meaning of situations, or bear responsibility for consequences. They may imitate language, behavior and analytical skills, or even simulate empathy and understanding, but they do not understand what they produce, for they lack the affective, relational and spiritual perspective through which human beings grow in wisdom. Even when these tools are described as capable of "learning," their way of doing so is different from that of a human person. It is not the experience of those who allow themselves to be shaped by life and grow over time through choices, mistakes, forgiveness and fidelity. Rather, it is a form of statistical adaptation based on data and feedback, which can be very effective, but does not imply inner growth.

A valuable tool that requires vigilance

100. In light of what has been said, we can better understand why AI can be a valuable tool and, at the same time, why it calls for a measured and vigilant approach. In recent years, its private use has expanded significantly, prompting growing reflection on both the opportunities it offers and the risks tied to its rapid spread. In personal use, three aspects in

particular deserve careful consideration: the ease with which results are obtained, the impression of objectivity and the simulation of human communication. The speed and simplicity with which information, complex analyses, media content and practical assistance can be accessed undoubtedly makes life easier. Yet they can also encourage excessive reliance and the search for ready-made answers, and weaken personal creativity and judgment. The apparent objectivity of the responses and suggestions these systems provide can lead us to overlook the fact that they reflect the cultural assumptions of those who designed and trained them, with all their strengths and limitations. The artificial imitation of positive human communication—words of advice, empathy, friendship and even love—can be engaging and at times genuinely helpful. However, for less discerning users, it can also be misleading, creating the illusion of a relationship with a real personal subject. When words are simulated, they do not build genuine relationships, but only their appearance. The artificial imitation of care or support can become particularly risky when it enters contexts where real relationships and emotional bonds are lacking. Here, the danger is not so much that a person may believe they are communicating with another person, but rather that they may gradually lose the very desire to form genuine human connections.

101. Broadening our perspective to the use of AI in society, we see that it is now embedded in decision-making processes across many sectors and at multiple levels: in

communication, management and control. The gains in efficiency and the potential to improve certain services are clear, yet rapidly and uncritically adopting them exposes us to a range of risks, including the tendency to overlook the environmental impact. Current AI systems require enormous amounts of energy and water, significantly influencing carbon dioxide emissions, and place heavy demands on natural resources. As their complexity increases, especially in the case of large language models, the need for computing power and storage capacity grows too, which requires an extensive network of machines, cables, data centers and energy-intensive infrastructure. For this reason, it is essential to develop more sustainable technological solutions that reduce environmental impact and help protect our common home.[124]

Responsibility, transparency and the governance of AI

102. The use of AI is never a purely technical matter: when it enters processes that affect people's lives, it touches on rights, opportunities, status and freedom. Important and sensitive decisions—concerning employment, credit, access to public services or even a person's reputation—risk being fully delegated to automated systems that do not know

---

124. Cf. Dicastery for the Doctrine of the Faith and Dicastery for Culture and Education, Note *Antiqua et Nova* (January 14, 2025), 96: *AAS* 117 (2025), 201.

"compassion, mercy, forgiveness, and above all, the hope that people are able to change,"[125] and can therefore give rise to new forms of exclusion. There are clearly harmful uses, such as the manipulation of information or violations of privacy. Yet there is also a subtler danger, for when AI systems present themselves as neutral and objective, they end up reflecting and reinforcing the stereotypes or ideological bias of their designers and developers.

103. Indeed, entrusting an algorithm in practice with the power to select who is worthy or not, without anyone bearing responsibility for that judgment, is to hand over the task of redefining the boundaries of human possibilities. In this process, political responsibility is also lost, not just empathy toward those excluded, which can, after all, be simulated. The exclusion of the vulnerable becomes cloaked in a veneer of neutrality and objectivity, against which it becomes difficult to raise objections. In this way, injustice goes unnoticed, and compassion, mercy and forgiveness—understood not as mere appearances but as real political actions—gradually disappear from view.

104. From this follows a simple but compelling consequence: we cannot consider AI to be morally neutral. In reality, every technical tool embodies choices and priorities through what it measures, ignores and optimizes, and how it

---

125. Francis, *Address to Participants at the Meeting of the "Minerva Dialogues" promoted by the Dicastery for Culture and Education* (March 27, 2023): *AAS* 115 (2023), 465.

classifies people and situations. If a system is designed or used in a way that treats some lives as less worthy, or excludes them without the possibility of appeal, then it is not merely a tool "to be used well," since it has already introduced criteria that contradict the inalienable dignity of the human person. For this reason, ethical discernment cannot be limited to asking whether we are using a system for good or bad purposes; it must also examine how that system is designed and what vision of the human person and society is embedded in the data and models that guide it.[126]

105. For AI to respect human dignity and truly serve the common good, responsibility must be clearly defined at every stage: from those who design and develop these systems to those who use them and rely on them for concrete decisions. In many cases, however, the internal processes leading to a result remain opaque, making it harder to assign responsibility and correct errors. This is where accountability becomes crucial: the possibility of identifying who must "account" for decisions, justify them, monitor them, and, when necessary, challenge them and remedy any harm caused.[127]

106. Calling for prudence, rigorous evaluation and even, at times, a slower pace in adopting AI does not mean opposing

126. Cf. Dicastery for the Doctrine of the Faith and Dicastery for Culture and Education, Note *Antiqua et Nova* (January 14, 2025), 41: *AAS* 117 (2025), 178.

127. Cf. ibid., 44–45: *AAS* 117 (2025), 179–180.

progress; instead, it is an exercise of responsible care for the human family. This need is all the more urgent given the frequent imbalance between the speed of technological growth and the slower development of awareness, norms, safeguards and institutions capable of governing its effects. It is not enough to invoke ethics in the abstract; robust legal frameworks, independent oversight, informed users and a political system that does not abdicate its responsibility are required. Otherwise, change will be governed only by technocratic thinking and presented as necessary and inevitable, ultimately imposing rules shaped by those who control data, infrastructure and computing power.

107. We cannot be satisfied with merely calling for the moralization of machines—the so-called "alignment" of AI with human values—without also having the courage to insist on a further condition: the possibility of openly discussing the ethical frameworks involved and subjecting them to shared standards of social justice. Otherwise, those who control AI will impose their own moral vision, which will become the invisible infrastructure of these systems. A more moral AI is not enough if that morality is determined by a few. What is needed is a more active political involvement that is capable of slowing things down when everything is accelerating, and of protecting the opportunities for communities still to be able to participate and ask questions.

108. In fact, as with every major technological shift, AI tends to amplify the power of those who already possess

economic resources, expertise and access to data. In light of the common good and the universal destination of goods, this raises serious concerns, since small but highly influential groups can shape information and consumption patterns, influence democratic processes and steer economic dynamics to their own advantage, undermining social justice and solidarity among peoples. For this reason, it is essential that the use of AI, especially when it touches on public goods and fundamental rights, be guided by clear criteria and effective oversight, grounded in participation and subsidiarity. Communities and intermediary organizations must not be reduced to passive recipients of decisions made elsewhere; they must be able to contribute to discernment and oversight. Moreover, ownership of data cannot be left solely in private hands but must be appropriately regulated. Data is the product of many contributors and should not be treated as something to be sold off or entrusted to a select few. It is necessary to think creatively in order to manage data as a common or shared good, in a spirit of participation, as Saint John Paul II already suggested regarding collective goods.[128]

109. The principles of Social Doctrine offer a framework for understanding this new reality. In a world where data, computational resources and regulatory influence remain in the hands of a few, to speak of the common good means

128. Cf. Saint John Paul II, Encyclical Letter *Centesimus Annus* (May 1, 1991), 40: *AAS* 83 (1991), 843.

exposing this new form of epistemic, economic and political asymmetry and naming the new monopolies of AI. To speak of the universal destination of goods means finding ways of ensuring universal access to both technologies and the education needed to use them. To speak of subsidiarity calls for protecting the ability of communities to make choices and corrections, rather than confining their role to mere oversight after the standards have been set elsewhere. To speak of solidarity obliges us to recognize the hidden, often exploited workers, who sustain algorithmic systems. To speak of justice requires questioning the global distribution of power that decides who in fact can train these models and who is merely subjected to them. Likewise, it means acknowledging that social justice is not only a goal to be safeguarded after technologies are deployed, but a condition that must shape their very design from the outset.

110. Finally, I would like to employ the expression "to disarm," which is close to my heart. Disarming AI means freeing it from the mentality of "armed" competition, which today is not limited simply to the military context, but is also an economic and cognitive phenomenon. This entails a race for ever more powerful algorithms and larger datasets, driven by the desire to secure geopolitical or commercial dominance. To disarm means discrediting the assumption that technical power automatically confers the right to govern. To disarm does not mean rejecting technology, but preventing it from dominating humanity. It means freeing technology from

monopolistic control and opening it to discussion and debate, therefore making it human-friendly and restoring it to the plurality of human cultures and ways of life. Our task today is not only ethical or technical. It is ecological in the deepest sense, for it concerns a new dimension of our common home. AI is already an environment in which we are immersed, as well as a force with which we must engage. For this reason, merely regulating it is insufficient; it must be disarmed, welcoming and accessible.

111. I wish to address a special appeal to those who develop artificial intelligence. In one sense, technological innovation can represent human participation in the divine act of creation. Developers, therefore, bear a particular ethical and spiritual responsibility, for every design choice reflects a vision of humanity. Just as the creator of an artistic or literary work must consider the values it conveys, so developers are called to embed values in their projects with due seriousness: with transparency, responsibility toward affected communities and careful attention to ensuring that what is being cultivated is a genuine good.

What must not be lost

112. Having considered the issues of responsibility and governance of AI, we must now return to our central question: what does it mean to safeguard our humanity? The risk extends beyond the misuse of certain technologies. More gravely, the

pervasive technocratic paradigm in which we are immersed, and that is amplified by the digital revolution and AI, threatens to normalize an anti-human vision. In that vision, the fullness of life is equated with having more, reducing weakness, eliminating uncertainty and exerting total control. When efficiency becomes the ultimate measure of value, human beings are tempted to see themselves as a project to be optimized rather than as persons called to relationship and communion.

113. In reality, elevating any single dimension of human existence to an absolute is always a mistake. Indeed, disorder does not arise only from scarcity; even unchecked growth can give rise to impoverishment. In an ecosystem, balance is disrupted when one species expands at the expense of others; in human life, something similar occurs when one faculty claims to be the measure of everything. Thus, intelligence, when absolutized, overshadows other essential dimensions of life, such as affection, the will, commitment and relationships. Similarly, technical power, if left unbalanced, does not make us more capable; it makes us more isolated and more vulnerable to being dominated and excluded. This critical point does not oppose intelligence, but serves as a reminder that when intelligence becomes self-referential, its true purpose of serving life and the human person is lost.

114. The quality of a civilization is measured not by the power of its means, but by the care it is able to offer, by its ability to recognize the other as a face not merely as a function. The ability to care for one another is a fundamental

dimension of our humanity, one that is learned and mastered through lived experience. Reading stories to a child, offering company to an elderly person and arranging a home so that it is welcoming, are simple gestures often rooted in family life. They teach us to value care at a societal level and train us to recognize others as persons worthy of attention. Technology can also support this mutual care between people, for example, by providing tools that help us anticipate and organize things, without undermining human freedom and judgment. After all, human beings are the subjects of relationships and responsible for their own decisions.

Underlying narratives: transhumanism and posthumanism

115. In an attempt to shed light on the cultural assumptions accompanying the ongoing digital revolution, I would now like to turn our attention to certain currents of thought that interpret progress as surpassing the human condition, and which are often grouped under the labels of transhumanism and posthumanism. These perspectives form the ideological background present in some centers of technological power and occupy the collective imagination in a simplified form, especially in the media and on social networks. They tend to foster enthusiasm for new technologies through a futuristic vision of an "enhanced human being" or "human-machine hybrid."

116. Transhumanism and posthumanism encompass a range of currents and sensibilities, making it difficult to define

them in a single, unambiguous way. They can be likened to an archipelago of conceptual "islands," distinct yet connected by a common "sea" of assumptions, namely the central role of technology and the aspiration to transcend the limits of the human condition. In general, transhumanism envisions the enhancement of human beings through technologies—such as biomedicine, body engineering, devices and algorithms—with the aim of increasing performance and capabilities. Posthumanism, especially in its more radical forms, goes further: it challenges anthropocentrism and envisions a hybridization of human beings, machines and the environment, even anticipating a threshold where humanity surpasses itself in a new evolutionary stage. Even when such ideas remain largely speculative, they gain relevance by altering the collective imagination and thereby influence social, economic and political choices.[129]

117. From the perspective of the Church's Social Doctrine, the key issue is not the use of technology as such, but the vision that underlies it. If the human being is treated as something to be perfected or surpassed, it becomes easier to accept that some lives are less useful, less desirable or less worthy. In the name of progress, "necessary sacrifices" may begin to be justified, placing the burden on the most vulnerable in pursuit

129. Cf. International Theological Commission, *Quo vadis, humanitas? Thinking about Christian anthropology in the face of some scenarios on the future of humanity* (February 9, 2026), 63.

of a supposed optimization of the species. In this regard, the aforementioned warning of Saint Paul VI retains great foresight: indeed, scientific and technological advances, when detached from moral and social progress, end up turning against humanity.[130] For this reason, a clear distinction must be made. It is one thing to integrate technology within a human-centered, relational vision; it is quite another to be guided by an outlook that devalues human limits and promises a purely technical form of "salvation."

The limit, the heart and the grandeur of the human person

118. Our relationship with life seems to be in crisis today. Everything that appears as a "limit"—incapacity, illness, old age, suffering, vulnerability—tends to be seen primarily as a defect to be corrected, rather than as a reality through which our humanity matures and opens itself to relationship. And yet we must remember that humanity flourishes not *despite* limitations, but often *through* them. The light of faith offers a perspective on reality that helps us recognize what we call the "contingency" of the things of this world. While it is right to strive to alleviate the suffering that marks human life, it is also wise to acknowledge our fundamental finitude, knowing that "religious experience, and in particular Christian faith,

---

130. Cf. Saint Paul VI, *Discourse on the occasion of the 25th Anniversary of the FAO* (November 16, 1970): *AAS* 62 (1970), 833.

propose that we live, without oversimplification, this ambivalence between human greatness and limitation, interpreting it in the light of our original and fundamental relationship with God."[131]

119. It is precisely within our limitations that the following find a place: compassion, as well as a sincere concern for the needs of others; a generosity that can emerge even in the midst of darkness and failure; spiritual experience and the worship of God. We see this at many moments when our limits become tangible: when we face rejection, when we suffer the illness or loss of a loved one, when we encounter our own weakness or failure. Mysteriously, it is precisely in such moments that we can discover a new wisdom, tangibly experience the closeness of others and encounter the presence of the Lord.

120. Even when limitations are experienced as inner suffering, human wisdom teaches us not to deny or suppress it, but to integrate it. To eliminate suffering entirely would mean, in the end, extinguishing love and desire as well. Those who love and desire cannot avoid passing through trial and suffering; and over the years, we carry within us lessons that leave their mark like scars, the memories of a journey shaped by freedom and failure, dreams and disappointments. It is only thanks to the interplay of these elements that the wonders of the soul

131. International Theological Commission, *Quo vadis, humanitas? Thinking about Christian anthropology in the face of some scenarios on the future of humanity* (February 9, 2026), 3.

occur within us, allowing us to sense the richness of our humanity.[132] To renounce this adventure, both tragic and splendid, in the name of a presumed transcendence of all limits, could mean many things, but it would no longer be human.

121. The moral corruption of our limitations as created beings—namely the evil that clearly agitates the human heart—ruins society and life, at times reaching extreme forms of inhumanity. Yet even these painful expressions of our limitations leave openings for the good. Even when persons dehumanize themselves and bring about tragedy, a small light continues to shine within humanity, one that can be rekindled, with God's grace, along paths of conversion and reconciliation. As Viktor Frankl rightly observed, in moments of horror, "we have come to know man as he really is. After all, man is that being who invented the gas chambers of Auschwitz; however, he is also that being who entered those gas chambers upright, with the Lord's Prayer or the *Shema Yisrael* on his lips."[133]

132. "If we devalue the heart, we also devalue what it means to speak from the heart, to act with the heart, to cultivate and heal the heart. If we fail to appreciate the specificity of the heart, we miss the messages that the mind alone cannot communicate; we miss out on the richness of our encounters with others; we miss out on poetry. We also lose track of history and our own past, since our real personal history is built with the heart. At the end of our lives, that alone will matter." Francis, Encyclical Letter *Dilexit Nos* (October 24, 2024), 11: *AAS* 116 (2024), 1372.

133. V. Frankl, *Man's Search for Meaning. An Introduction to Logotherapy*, Boston 1963, 213.

122. Finitude, when truly accepted, does not diminish us but opens us to recognizing the face of God and others. Indeed, precisely because we experience limits—vulnerability, suffering and failure—we can recognize the inviolable dignity of every person, both our own and that of others. In this same experience, we remain capable of intuiting a fraternity greater than ourselves and of perceiving injustice as a scandal. Authentic culture and art preserve this spark, resisting the normalization of evil. For this reason, certain works have taken on an almost prophetic significance: Beethoven's Ninth Symphony can be seen as a desire for unity; *Guernica* as a denunciation of dehumanization; *Schindler's List* as a call not to consign the past to oblivion.

123. History does not appear solely as a record of human violence, but also as evidence that humanity is capable of creating institutions that protect our shared life. Over the past two centuries, this can be seen in several emblematic achievements: the founding of the International Committee of the Red Cross (1863), whose operational neutrality ensures compassionate care for all; the long process that led to the abolition of slavery, which represented not only a legal shift but a transformation of conscience; the establishment of the United Nations (1945) and the *Universal Declaration of Human Rights* (1948), which articulated a shared language for affirming, at least as a common ideal, the universality of human dignity; and the *1951 Refugee Convention*, which recognizes the duty to protect those fleeing persecution and danger. In

each of these cases, the desire for good took concrete shape in public contexts—laws, institutions and practices—capable of limiting the abuse of power and defending the vulnerable. Yet none of these developments emerged without encountering resistance, narrow interests or cultural inertia. Moral progress almost always unfolds through a long and demanding journey, often marked by setbacks. We need only think of stalled peace processes or the slow implementation of environmental commitments. The very fragility of these achievements highlights how precious the responsibility is of those who initiate and sustain them.

124. Certain events make it clear that history can also change when individuals truly take the dignity of everyone seriously: the civil rights movement in the United States of America, closely associated with the testimony of Martin Luther King Jr., or the end of apartheid in South Africa following the release of Nelson Mandela and his decision not to surrender the future to hatred. In different contexts, many courageous and generous women have also stood out, including Saint Laura Montoya, Saint Teresa of Calcutta, Dorothy Day, Marie Skłodowska-Curie, Maria Montessori, Elisabeth Elliot, Wangari Maathai, Benazir Bhutto and countless others from every continent whose commitment has contributed to making history more humane.

125. Alongside these public signs, there is a more hidden but decisive story. We see it in religious communities that choose to serve in poor and dangerous places. We also see it

in the martyrs of fraternity and justice, such as Saint Maximilian Mary Kolbe, Saint Oscar Romero and Blessed Enrique Angelelli; and in those witnesses who embodied the hope of the Gospel as well as human dignity amidst harsh, often inhumane conditions, such as Venerable Francis-Xavier Nguyễn Văn Thuận. Above all, it is visible in the "martyrs of everyday life" who care for, educate, accompany and comfort without fanfare, such as parents, nurses, doctors, volunteers and those who remain alongside an elderly person or an outcast. Their testimony demonstrates that goodness does not advance automatically, but requires the perseverance, memory and interior conversion necessary to begin anew, even after defeat.

126. It is this intertwining of just institutions, credible witnesses and daily fidelity that sustains hope and provides clear direction for technological progress without allowing the heart to regress. For this reason, humanity —in all its grandeur and woundedness—must never be replaced or surpassed. We can embrace the technological progress that alleviates suffering and unlocks new possibilities, provided that we do not abandon the very essence of our humanity, namely the capacity for relationship and love. This leads to a crucial question: if an authentic "more than human" exists, where is it to be found? The Christian faith answers that question by pointing to a fulfilment that does not arise from a technological divinization, but through God's grace received in Christ.

The authentic "more than human": grace and Christian humanism

127. The expression "more than human" is not an exclusive domain of technological promise. For centuries, the Christian tradition has maintained that human beings are not confined by the boundaries of their own nature; rather, they are called to self-transcendence, not through an escape from reality or a contempt for their limitations, but through their fulfillment in love. Faith recognizes an openness toward the "beyond," which originates as a gift from God. This transformation is a work of the Holy Spirit. As Saint Thomas Aquinas taught, this process of elevation and transformation "surpasses every capability of created nature,"[134] for an infinite disparity separates our finite nature from the life of God.[135] Nevertheless, it remains possible to enter into the heart of that inexhaustible life, even as we journey through the limitations of this world. The one who makes this passage possible can only be the Eternal One who gives of himself. Indeed, it is God himself who overcomes the "infinite" disproportion.[136] In him, the re-creation of the human person happens. "If anyone is in

---

134. Saint Thomas Aquinas, *Summa Theologiae*, I–II, q. 112, a. 1, co; q. 114, a, 5, co.: ed. Leonina, VII, Rome 1892, 323 and 349.

135. Cf. ibid., q. 114, a. 1, co.: ed. Leonina, VII, 344.

136. Cf. Saint Thomas Aquinas, *Super Boetium de Trinitate*, q. 1, a. 2, ad 3: ed. Leonina, L, Rome 1992, 96; *Summa Theologiae*, I, q. 7, a. 1, ad 3: ed. Leonina, IV, Rome 1888, 72.

Christ, there is a new creation: everything old has passed away; see, everything has become new" (2 Cor 5:17).

128. When we embrace the possibility of transcending ourselves through God's grace, we do not deny our nature, nor do we become less human. On the contrary, as Pope Francis explained, "We become fully human when we become more than human, when we let God bring us beyond ourselves in order to attain the fullest truth of our being."[137] Herein lies the radical departure from Promethean dreams: what saves humanity is not enhanced self-sufficiency, but a relationship that liberates, a communion that transforms. In this light, a technology that merely classifies and optimizes what already exists can, however unintentionally, become an obstacle to change and growth. For an algorithm, an error is a flaw to be corrected; for a person, however, an error can be a catalyst for profound change. A person's future is not calculable, but depends on one's freedom—elevated by the inexhaustible grace of God—and on the relationships cultivated.

Two cities and two loves

129. Christian humanism does not reject science or technology, but embraces them with gratitude and realism, and grounds them within a higher vocation. The creative intelli-

137. Francis, Apostolic Exhortation *Evangelii Gaudium* (November 24, 2013), 8: *AAS* 105 (2013), 1022.

gence of humanity is a gift that can alleviate suffering and open up new possibilities, but it must remain ordered toward the common good, justice, the care of the vulnerable and creation. In this sense, the true alternative is not between enthusiasm and fear, but between two paths of development: a progress that serves individuals and peoples, or a progress that subjects them to the mentality of power. Ultimately, the key question remains the one posed by Saint John Paul II: does AI "make human life on earth 'more human' in every aspect of that life? Does it make it more worthy of man?"[138] If the answer is yes, then we can recognize it as an opportunity to be embraced responsibly, on a path of patient, shared reconstruction, akin to the rebuilding of Jerusalem narrated in the Book of Nehemiah. If, however, power grows while the heart withers and human bonds fray, then we are faced with a new form of Babel—a construction that is grandiose, yet fundamentally dehumanizing.

130. Questioning this alternative path of progress and how we interpret and live it is ultimately a matter of examining our own hearts. The way we understand and shape relationships, work and institutions, in practice reveals our fundamental values. In the end, it all stems from what we hold most dear. This is a love that guides us as to what we truly cherish, both as individuals and as a society, and directs

138. Saint John Paul II, Encyclical Letter *Redemptor Hominis* (March 4, 1979), 15: *AAS* 71 (1979), 286–287.

our lives and actions. Saint Augustine described human history as a struggle between two loves, which give rise to two ways of inhabiting the world and living together—or two "cities," as it were: on the one hand, the love of God and neighbor; on the other, the exclusive love of self. "Two loves have built two cities: the earthly city, the love of self even to the contempt of God; the heavenly city, the love of God even to the contempt of self."[139] As throughout history, these two loves continue to contend for dominance in our hearts today. The age of AI is no exception: the construction of Babel or the rebuilding of Jerusalem begins within each one of us.

139. Saint Augustine, *De civitate Dei*, XIV, 28: *CCSL* 48, Turnhout 1955, 451.

Chapter Four

Safeguarding Humanity at a Time of Transformation

Truth, Work, Freedom

131. Having outlined the context in which the challenge of technological transformation is situated, especially those linked to AI and to transhumanist and posthumanist currents, we cannot remain at the level of general analysis alone. When languages and tools change, so do everyday actions and social relationships. For this reason, we must focus on certain areas in which these transformations have particularly concrete, and at times tragic, consequences. In light of the principles of the Church's Social Doctrine, the digital transformation invites us to rediscover truth as a common good, to protect the dignity of work and to safeguard freedom against all forms of dependence and commercialization.

Truth as a common good

Truth and democracy

132. The use of digital platforms and AI systems is driving profound changes in public and political communication. Tools that could foster dialogue and participation are often used to construct distorted narratives and blur the boundaries between truth and falsehood, mixing facts with opinions. Disinformation did not begin with AI, yet today it finds a powerful amplifier in AI. The ability to manipulate content, images and videos exposes people to biased or misleading perspectives. This problem has both cultural and moral dimensions, since the quality of public communication depends directly on social trust and, in turn, shapes it. At the same time, truthful information does not arise from centralized or automated control. In public discourse, the truth of facts has a rational dimension, as it requires verification, cross-checking of sources and responsible argumentation. Moreover, it is deeply relational, built through bonds of trust and shared practices, as well as an honest exchange with others and with the world. Only the shared pursuit of the veracity of facts, perceived as a common good, can provide a solid foundation for just communication.

133. Those who command powerful technological and economic resources, along with substantial human capital for intervention, possess significant capabilities for influencing cultural change. Ultimately, they can influence a significant

number of people concerning the truth about humanity, the world, the meaning of existence, the family and even God. This is pure power detached from truth, which subtly or overtly imposes what it wishes others to accept as true. At its root lies a deeper and often unrecognized "sickness": the fact that "modern man is wrongly convinced that he is the sole author of himself, his life and society. This is a presumption that follows from being selfishly closed in upon himself."[140] Consequently, people believe that they can construct reality, and that whatever best suits their claims corresponds to what is true. Saint John Paul II reflected on the consequences of this "crisis of truth," going so far as to state that "once the idea of a universal truth about the good, knowable by human reason, is lost, inevitably the notion of conscience also changes."[141] In such a context, universally valid truths, which precede us and which conscience must accept, are no longer recognized. This led Pope Francis to ask with realism: "What is law without the conviction, born of age-old reflection and great wisdom, that each human being is sacred and inviolable?" To which he concluded: "If society is to have a future, it must respect the truth of our human dignity and submit to that truth. Murder is not wrong simply because it is socially

140. Benedict XVI, Encyclical Letter *Caritas in Veritate* (June 29, 2009), 34: *AAS* 101 (2009), 668–669.

141. Saint John Paul II, Encyclical Letter *Veritatis Splendor* (August 6, 1993), 32: *AAS* 85 (1993), 1159.

unacceptable and punished by law, but because of a deeper conviction. This is a non-negotiable truth attained by the use of reason and accepted in conscience. A society is noble and decent, not least for its support of the pursuit of truth and its adherence to the most basic of truths."[142]

134. The search for truth is an essential element of democracy, which is itself a means of contributing to the common good. When questions about what is true lose their appeal, and a pragmatism takes hold that is content with what appears useful or effective, then democratic life is weakened. After all, democracy does not consist of rules and procedures alone, but above all of a solid concordance with the facts and a genuine commitment to the good of individuals and society as a whole. Indifference to the truth leads, slowly but surely, to a descent into totalitarianism. As the philosopher Hannah Arendt wrote, the ideal subjects of such regimes are not so much those who are ideologically convinced, but rather "people for whom the distinction between fact and fiction (i.e., the reality of experience) and the distinction between true and false (i.e., the standards of thought) no longer exist."[143]

Communication and the collective imagination

135. In view of this, it is important to recall that communication "is not only the transmission of information, but it is

---

142. Francis, Encyclical Letter *Fratelli Tutti* (October 3, 2020), 207: *AAS* 112 (2020), 1043.

143. H. Arendt, *The Origins of Totalitarianism*, III, New York 1962, 474.

also the creation of a culture."[144] The content that circulates within digital environments shapes how people perceive the world and introduces into the collective consciousness images and narratives that direct our desires and influence our daily choices. This is "not a parallel or purely virtual world,"[145] since what originates online now becomes a part of people's lives, especially of the youngest.

136. For this reason, those who control digital platforms and means of communication have a considerable ability to affect the collective imagination and to present a particular vision of reality as desirable. Such power should be constantly guided by the pursuit of truth and respect for human dignity, so that the culture fostered on the internet does not become an instrument of excessive distraction, homogenization or dominance, but rather a setting in which inner freedom and critical thought can mature.

Toward an ecology of communication

137. Our first task is neither to demonize nor idolize technological tools, but to utilize them on the basis of a fundamental principle, namely that truth is a common good and not the property of those with power or influence. We must

---

144. Leo XIV, *Address to Representatives of the Media* (May 12, 2025): *AAS* 117 (2025), 681–682.

145. Benedict XVI, *Message for the 47th World Day of Social Communications* (January 24, 2013): *AAS* 105 (2013), 183.

therefore promote an ecology of communication. On the level of public policy, this entails establishing norms so that the decision-making behind content selection and its development becomes more transparent and protects personal data. Regarding social and cultural aspects, this requires a strengthening of intermediary organizations, serious journalism and forums for debate, where reasoned argumentation and verification carry greater weight than immediate reaction. For families and schools, there is a growing need for new educational awareness and for formation concerning the proper and critical use of digital tools, AI and online commercial and financial platforms. In universities, the principal challenge lies in the integration of knowledge, cultivating both the capacity to connect and synthesize knowledge in order to grasp complexity, and the skills necessary to verify facts.

138. Christian communities, too, are called to commit themselves to transparency in communication and to the honest pursuit of facts. Sadly, this has not always been the case. We have witnessed with shame the emergence of painful truths concerning even members of the Church and ecclesial realities. In particular, some journalists, driven by a passion for truth, have played a crucial role in bringing injustices and abuses to light. To them, I wish to repeat the words that Pope Francis used in speaking to journalists: "I also thank you for what you tell us about what goes wrong in the Church, for helping us not to sweep it under the carpet, and for the voice

you have given to the victims of abuse."[146] Yet vigilance and transparency remain first and foremost a grave responsibility for the Church herself, and we must not wait for others to compel us to confront uncomfortable truths about ourselves.

An educational alliance for the digital age

139. In an era when truth is often distorted in order to serve particular interests and communication strategies, the field of education assumes decisive importance. Yet rapid technological transformations reveal just how unprepared we are on the educational level. The pervasiveness of digital media fosters a culture of immediacy and hyper-stimulation, which gives rise to fatigue, boredom and apathy concerning the effort required for seeking the truth.

140. Education, by contrast, is a long journey requiring patience, and therefore needs time for development and for engagement with reality beyond appearances. This is a fundamental issue because every technology shapes those who use it. Educating people about the use of AI, then, involves teaching them to decide when and for what purpose it ought *not*

146. Francis, *Address on the occasion of the Conferral of the rank of Knight and Dame of the Grand Cross of the Pian Order to Mr Philip Pullella and Ms Valentina Alazraki* (November 13, 2021): *L'Osservatore Romano*, November 13, 2021, 12.

to be used. The speed and ease with which answers or summaries can be obtained risk extinguishing the desire to ask questions, which is a process that bears fruit only over time. As Plato wrote, the deepest and most important things are learned only after much time and effort, by engaging in discussion with others, "striking upon" ideas and experiences together like flint until the spark of understanding is kindled within us.[147] We must learn, then, how to exercise restraint in the use of AI and to protect our young people from the promise of the perfect machine, from that subtle temptation which renders human thought seemingly superfluous precisely when it is most needed.

141. In recent years, psychological and psychiatric literature has documented with growing insistence how early and unsupervised exposure to digital devices and social media can negatively impact sleep, attention span, control of emotions and relationships, especially during the most vulnerable stages of life, at times with tragic consequences. This is further aggravated by easy access to violent or degrading content that offends sensibility, to pornographic and hypersexualized material, to messages that trivialize the body and emotions, and to proposals that normalize risky behavior. Online phenomena such as grooming, blackmail and the sexual exploitation of minors are not uncommon, and are made more

147. Cf. Plato, *Letter VII*, 344b–c: ed. Souilhé, XIII/1, Paris 1931 (*CUF, Série grecque* 63), 54.

insidious by the use of fake profiles, algorithms that facilitate dangerous contact, and AI tools capable of manipulating images and videos. Having a personal mobile device at too early an age and using it without adult supervision can exacerbate young people's vulnerabilities, foster addiction and expose them to isolation, bullying and cyberbullying, as well as to pressures to share intimate images or sensitive information.

142. It is difficult for parents by themselves to resist the influence of business models that monetize attention and time. Therefore, it is essential to form an alliance among policy-makers, educational institutions and families that is capable of concretely supporting adults in this task. Far-sighted public policies are needed to oppose the immediate interests of platforms, concentrated in a few hands, when they conflict with the wellbeing of minors. In this regard, interventions by legislators are appropriate for setting age limits, holding service providers accountable rather than shifting the whole burden of control onto families, and for providing specific protections against all forms of online sexual exploitation and violence. Thus can children and adolescents, who are entrusted to our care, be genuinely protected as a precious treasure.[148] At the same time, it is also necessary to teach

---

148. Cf. Leo XIV, *Address to the Participants in the Conference "The Dignity of Children and Adolescents in the Age of Artificial Intelligence"* (November 13, 2025): *L'Osservatore Romano*, November 13, 2025, 3.

children, adolescents and young people how to recognize manipulation, defend their dignity and respect that of others in digital environments.[149]

The central role of schools

143. School is the place where new generations can learn to seek and love the truth, to reflect on the meaning of life and to recognize the dignity of every person. For this reason, many parents, who want their children to grow in the capacity to form relationships, develop critical thinking skills and embrace solid values, place great expectations on schools as valuable partners in their children's education. Yet parents have the primary and inalienable right to choose the kind of education and formation for their children, in a manner consistent with their moral, cultural and religious convictions. Today, the world of education faces a number of urgent challenges.

144. The first challenge is socio-political. Both within individual nations and across different regions of the world, significant inequalities persist concerning access to basic education and higher studies. In many nations, Governments have not yet invested the necessary resources for guaranteeing a quality education for all, whether by adequately supporting

149. Cf. Leo XIV, *Address to the members of the Advisory Board of the RCS Academy* (November 7, 2025): *L'Osservatore Romano,* November 7, 2025, 4.

the public school system or by assisting private institutions that offer this essential service. When a substantial portion of education, at various levels, is entrusted to private institutions, access to schooling may become overly dependent on families' financial means, especially in the absence of adequate public support. In the face of this risk, it is nevertheless important to acknowledge and encourage the contribution of the many private Catholic educational institutions which ensure inclusive access for children and young people of every background, even when families' economic circumstances would not otherwise allow it.

145. The second major challenge is pedagogical. Many educational systems struggle to keep pace with change and to support the integral development of students. The advance of information technologies and AI is rapidly rendering curricula obsolete that were designed for a different era. Meanwhile, the organization of schools, physical spaces, evaluation methods and the role of teachers themselves must be rethought in order to promote an authentically integral education that addresses every dimension of the person. It is necessary to support the ongoing formation of teachers throughout their professional lives, so that they can engage positively with new technologies, helping students to use them responsibly, critically and creatively, rather than passively succumbing to their influence.

146. The third major challenge is intellectual and concerns knowledge. Without careful attention, an educational system

lacking in a love for truth may emerge, in which an incessant flow of information replaces the essential exercise of research, reflection and discernment. As knowledge becomes increasingly fragmented, it becomes difficult to grasp reality as a whole, to ask profound questions about meaning, or to develop authentic, critical and creative thought. Many educators already report signs of dehumanization, where people may "know many things" but struggle to find direction in their lives, partly due to an inability to connect information with deeper knowledge or maintain a sense of purpose. A genuinely healthy attitude is needed, requiring rhythms that incorporate silence, in-depth study, reading and judicious analysis, for without these elements inner freedom may be compromised.

147. The Church's Social Doctrine invites families, schools, Christian communities and public institutions to form a renewed educational alliance. This takes shape when fundamental principles are translated into educational goals, including teaching students a sense of moderation and limits; recognition of the rights of others and of future generations to enjoy the goods that are either provided for us or made available by human ingenuity; freedom and responsibility; and a sense of transcendence and the common good. Schools are not called to follow the pace of the digital world, but to offer that which the digital sphere by itself cannot provide, namely a shared time for learning and developing trustworthy relationships.

The dignity of work at a time of digital transition

The value of work

148. Since the emergence of her Social Doctrine, beginning with *Rerum Novarum*, the Church has emphasized the protection of workers and the need to combat all forms of exploitation. Above all, however, the Magisterium has recognized in work "the essential key"[150] to understanding the entire social question, since it is through their work that individuals develop many dimensions of their existence. In view of this, we can understand the great intuition of Saint Benedict of Nursia, who united prayer and work, showing daily activity to be a part of the human response to God's call. Created in the image of the Creator, our own work in some way continues his, for thereby we contribute to the progress of society and the common good, put to good use the capabilities we have received, improve and beautify the world, support our families, engage in cooperative relationships and, through listening and dialogue, learn to build together something that no one could achieve alone.

149. For these reasons, work is not simply an instrument; it expresses and enhances the dignity of our lives. It is a

---

150. Saint John Paul II, Encyclical Letter *Laborem Exercens* (September 14, 1981), 3: *AAS* 73 (1981), 584.

requirement of the human condition, a normal path toward maturity, development and personal fulfilment. In this regard, financial assistance to the poor may at times be necessary in emergencies, but it cannot become the sole response, since the goal is to enable each person to live with dignity through his or her own work.[151]

150. Today, the convergence of automation, robotics and AI is rapidly transforming the very structure of work. It is said that this will bring great improvements for everyone. In reality, however, the "new ways" of working are not necessarily better, for "while AI promises to boost productivity by taking over mundane tasks, it frequently forces workers to adapt to the speed and demands of machines, rather than machines being designed to support those who work. As a result, contrary to the advertised benefits of AI, current approaches to technology can paradoxically de-skill workers, subject them to automated surveillance and relegate them to rigid and repetitive tasks. The need to keep up with the pace of technology can erode workers' sense of *agency* and stifle the innovative abilities they are expected to bring to their work."[152] Precisely in order to avoid this drift, it is

---

151. Cf. Francis, Encyclical Letter *Laudato Si'* (May 24, 2015), 128: *AAS* 107 (2015), 898.

152. Dicastery for the Doctrine of the Faith and Dicastery for Culture and Education, Note *Antiqua et Nova* (January 14, 2025), 67: *AAS* 117 (2025), 188–189.

necessary to design systems that are centered on the human person and not solely on performance.

The problem of unemployment

151. Saint John Paul II recognized that unemployment is a grave evil. Indeed, when it reaches massive proportions, it becomes a true social calamity that especially requires the State to exercise responsibility.[153] Today, amid the "fourth industrial revolution," this concern is even more acute, as innovation is often pursued solely for reducing costs and increasing profits.[154] In some contexts, there is a legitimate fear of a significant and rapid contraction in available jobs that would create a chain reaction deeply impacting families, young people and local economies. In many sectors, this can already be seen in new forms of job insecurity and inequality, characterized by outsized remuneration for a highly specialized minority alongside declining wages for a large portion of the workforce.

152. It is certainly desirable for technology to relieve humans of arduous, repetitive or dangerous tasks and to provide intelligent support for human activity. Yet, the protection

---

153. Cf. Saint John Paul II, Encyclical Letter *Laborem Exercens* (September 14, 1981), 18: *AAS* 73 (1981), 622–625.

154. Cf. Francis, Encyclical Letter *Laudato Si'* (May 24, 2015), 109: *AAS* 107 (2015), 891.

of employment opportunities and the irreplaceable role of the individual must remain the general rule. The pursuit of greater profits cannot justify choices that systematically sacrifice jobs, because the human person is an end, not a means, and the economic order must remain subordinate to human dignity and the common good.

153. At the same time, we must acknowledge that every real transition involves discontinuities, for it is uneven, fragmented and sometimes conflictual. Consequently, no single model of change or universal solution exists, since there are places and situations that require different responses. Given the inequality that characterizes our world, the spread of AI and computational systems produces varied effects in different places. Wealthy societies automate rapidly and chaotically, reducing the need for a workforce and creating room for unemployment and institutional friction. Vast regions of the world, by contrast, remain trapped in hybrid economies, where underpaid human labor and partial technologies coexist without achieving genuine transformation. These areas become places of precarious labor, and hotbeds of instability and forced migration. Therefore, solutions must be sought at national and local levels through the involvement of intermediary communities. We need adaptive tools, including well-structured models, local initiatives, progressive redistribution and new rights of access to essential goods. While not pursuing an abstract harmony, we must build concrete forms of human coexistence at this time of transformation.

154. Work remains a fundamental dimension of the human experience, for not only is it a means of sustenance, but it is also a context for expression, relationships and contributing to the community. Therefore, the problems related to work extend beyond the income necessary for family survival. A society that guarantees employment to only a small fraction of the population, despite having a high level of technical development, risks exposing many to forced inactivity, a lack of responsibility and the absence of daily tasks and stimuli, resulting in human and cultural impoverishment. This creates a paradox of material progress and anthropological regression that undermines the foundations of a just and stable social peace. For this reason, the Church's Social Doctrine insists that access to work for all must be a high priority for public policies and economic processes, serving as a criterion for evaluating the human quality of any development model.[155] Moreover, in those parts of the world where work tends to diminish or change radically due to technological and organizational processes outside of democratic control, we must rethink the nature of work and its connection to citizenship, ensuring that unemployment does not jeopardize social participation.

155. In light of this conviction, we can better appreciate the history of the Church's Social Doctrine after *Rerum Novarum*. The initiatives which emerged from that tradition,

---

155. Cf. Benedict XVI, Encyclical Letter *Caritas in Veritate* (June 29, 2009), 32: *AAS* 101 (2009), 666.

including associations, trade unions, cooperatives and welfare organizations, have contributed decisively to improving labor legislation, protecting the most vulnerable and promoting more humane conditions.[156] Today, however, these instruments are no longer sufficient by themselves in the face of the transformations driven by AI, the new organization of markets and the competitiveness that is rarely concerned with social sustainability. New collaborative efforts are needed among political leaders, labor organizations, the business world and the scientific community in order to develop rapidly adequate shared regulations and protections, including at the international level.[157] Labor unions, which the Church has consistently supported, are called upon to be open to new types of employment and the corresponding needs of workers, in order to represent and defend them. In this context, without bold decisions, the prospect of greater poverty and inequality looms large, which would leave many individuals marginalized, stranded and surrounded by the machines and automated systems that have replaced them.

156. At this time of transition, it is not enough to react only when jobs disappear; we must oversee the transformation in advance. One viable path is, first of all, to establish social criteria for innovation. Here, every introduction of automation and

---

156. Cf. Pontifical Council for Justice and Peace, *Compendium of the Social Doctrine of the Church*, 268.

157. Cf. Benedict XVI, Encyclical Letter *Caritas in Veritate* (June 29, 2009), 64: *AAS* 101 (2009), 698.

AI should be accompanied by verifiable measures to protect the employment, retraining and participation of workers. In this way, technology will be oriented toward freeing up human time and capabilities, rather than producing exclusion. Second, we need proactive policies that make continuous training and professional transitions accessible to all, ensuring that the cost of adaptation does not fall solely on individuals. Finally, there needs to be a corporate commitment to include quality and dignity of work among its indicators of success. When these conditions are present, innovation can serve as an ally of safer, more creative and dignified work; without them, innovation tends to become an accelerator of injustice.

An economy that values dignity

157. The labor market is one area in which the risks associated with new technologies more clearly emerge. It is thus necessary to remember that economic freedom is not absolute; it must always be measured against the common good and the dignity of every person. Entrepreneurial initiative can indeed be a true vocation, generating wealth and improving lives, rather than a variable that is dependent only on profit. This is possible when it recognizes that the creation of dignified, valuable jobs are an essential part of its proper service to society.[158]

158. Cf. Francis, Encyclical Letter *Laudato Si'* (May 24, 2015), 129: *AAS* 107 (2015), 899.

158. With prophetic spirit, Pope Francis warned against an economic freedom proclaimed in words alone, while actual conditions prevent many from benefiting from it.[159] Economic models that exalt efficiency and individual success often view investment in disadvantaged people or in those with slower development paths as useless or inconvenient, as if their futures depended solely on their ability to keep pace with the "winners." In reality, a just society requires a vigilant State and civil institutions that are capable of overcoming the singular mentality of efficiency, and of ensuring that resources, creative solutions and regulations favor the most vulnerable.[160] Instead of waiting for the benefits of growth to reach the poor "eventually," decisions need to be taken to ensure that growth becomes inclusive from the outset. The experience of recent decades shows that in economic and financial crises, it is always the poor who pay the highest price, while the theories that promise automatic general prosperity often prove to be illusory.

159. It is important to move beyond the current metrics of development—which for more than eighty years have been tied to the concept of Gross Domestic Product (GDP) —since these metrics almost systematically neglect aspects essential to the overall wellbeing of people and the environment. The development of parameters and metrics comple-

159. Cf. ibid.

160. Cf. Francis, Encyclical Letter *Fratelli Tutti* (October 3, 2020), 108: *AAS* 112 (2020), 1006.

mentary to GDP is crucial for improving the databases used for conducting analyses, political and economic decision-making and establishing regional, national and international priorities. The introduction of new parameters will allow for a comprehensive and timely assessment of how legislative and regulatory decisions impact the dignity of work, shared prosperity, inequality reduction and environmental protection. It will also affect the concept of development, educational processes, mindsets and public opinion, as well as peace, which is only authentic when based on justice.

160. In recent years, finance has increased in importance and has undergone significant innovation, driven partly by the introduction of cryptocurrencies. The reflections and observations contained in the teaching of my predecessors, particularly in their Encyclicals, have highlighted how the financial intermediation sector, "when operating without the necessary anthropological and moral foundations, has not only produced manifest abuses and injustice, but also demonstrated a capacity to create systemic and worldwide economic crisis."[161] It is likewise the case that income from capital risks replacing income from labor, which is often confined to the margins of the economic system's primary interests. Yet savings

161. Cf. Dicastery for the Doctrine of the Faith and Dicastery for the Promotion of Integral Human Development, *Oeconomicae et Pecuniariae Quaestiones. Considerations for an Ethical Discernment Regarding some Aspects of the Present Economic-Financial System* (January 6, 2018), 6: *AAS* 110 (2018), 772.

transformed into credit for the real economy, thereby creating both jobs and self-employed work, remain central for development and the investments that must accompany ongoing transitions. The social function of credit remains irreplaceable. Finance for its own sake is fundamentally different from finance aimed at the development, creation and evolution of work.

161. This perspective needs to become part of a broader view of global dynamics. While the world's wealth has grown in absolute terms, it is increasingly concentrated in fewer hands, widening inequalities both within and between countries. "There are a few who have too much, and too many who have little, that is the logic of today."[162] Scientific and technological advances, even in the medical field, are not easily accessible to the vast majority of people, as was dramatically demonstrated during the recent pandemic. While some regions spend heavily on superfluous interventions or dreams of individual enhancement accessible only to a select few, other parts of the world lack the essential equipment needed to save millions of human lives. To think that new technologies will automatically benefit everyone is to ignore the evidence. Unless transformations at the design stage prioritize

---

162. Francis, *Greeting to the staff of the International Fund for Agricultural Development* (*IFAD*) (February 14, 2019): *AAS* 111 (2019), 309. Cfr. Benedict XVI, Encyclical Letter *Caritas in Veritate* (June 29, 2009), 22: *AAS* 101 (2009), 657.

the prevention of new and further disparities, technological progress will inevitably produce structural inequalities. Today, justice requires access to the benefits of innovation, including care, knowledge, tools and opportunities.

162. Just laws and methods of redistribution are certainly necessary for correcting imbalances, including tax systems that lighten the burden on the weakest and ask for more from those with greater resources. However, the pursuit of social justice should not be considered a separate issue that follows only after the production of wealth, as if the economy existed solely to create wealth, with politicians only intervening afterwards in order to distribute it. Indeed, justice concerns every phase of economic activity, from resource acquisition to financing, and from production to consumption; every choice has moral consequences.[163]

163. More than ever, in the age of AI and robotics, it is no longer possible to rely solely on the "invisible hand" of the market.[164] Politics has the task of orientating economies and technologies to the common good, promoting dignified work, social inclusion and an equitable distribution of the benefits of innovation. Since many economic decisions transcend national borders, there is also a need for international cooperation capable of defining common strategies, especially in favor of

163. Cf. ibid., 36: *AAS* 101 (2009), 671–672.

164. Cf. Francis, Apostolic Exhortation *Evangelii Gaudium* (November 24, 2013), 204: *AAS* 105 (2013), 1105–1106.

the most vulnerable countries and people, in order to promote development and overcome welfare dependency. The thinking behind these choices is the immeasurable dignity of every person, the common good and a world truly governed for everyone. The interdependence between peace and development, as Saint Paul VI prophetically wrote in 1967,[165] remains applicable today, for prosperity contributes to building and reinforcing peace only if it is widespread, inclusive and sustainable.

164. In practical terms, in the age of AI and robotics, ensuring that the economy favors human dignity means adopting certain criteria for firm action. First, transparency and accountability: when data and algorithms influence credit distribution, personnel selection or access to services and opportunities, it is necessary that decisions be understandable, contestable and subject to oversight, so that individuals are not reduced to mere profiles. Second, inclusion and access: the benefits of innovation must be paired with investments in skills, infrastructure and essential services to ensure that technology does not widen the gap between those who have and those who have not. Finally, measures to ensure equity: taxation, social protection and industrial policies must correct the imbalances created by the concentration of wealth and power. Indeed, these criteria do not constitute a curb on innovation; instead they make it civilized and humane.

---

165. Cf. Saint Paul VI, Encyclical Letter *Populorum Progressio* (March 26, 1967), 87: *AAS* 59 (1967), 299.

Families and young people: the social conditions for hope

165. The family is a primary social good. Founded on the enduring union between a man and a woman, it is the first environment in which all persons develop their potential, become aware of their dignity and learn the earliest forms of truth and goodness, internalizing the habits that prepare them for life in society.[166] As the first natural society, endowed with foundational rights, the family is the fundamental and irreplaceable cell of every community organization.[167] Consequently, when political projects and major economic decisions relegate the family to a marginal or secondary role, the authentic growth of the entire social body is compromised.[168]

166. The family, however, is a fragile social good immediately affected by the economic and technological transformations reshaping the nature of work. It thus requires cultural, juridical and economic support. The devastating impact of unemployment and job insecurity on family structures is well known. In the short term, it may seem advantageous to reduce labor costs or maximize financial efficiency, but in the long

---

166. Cf. Saint John Paul II, Encyclical Letter *Centesimus Annus* (May 1, 1991), 39: *AAS* 83 (1991), 841.

167. Cf. Pontifical Council for Justice and Peace, *Compendium of the Social Doctrine of the Church*, 211.

168. Cf. Saint John Paul II, Letter to Families *Gratissimam Sane* (February 2, 1994), 17: *AAS* 86 (1994), 903–906.

term this undermines the very foundations of social coexistence. While technological successes are celebrated, the social fabric is progressively eroded, as if by a silent virus.

167. For young people, job insecurity is particularly devastating. As the Bishops of the United States of America have recalled, work is not merely a source of income but a crucial sphere in which identity is formed, friendships and relationships are forged, practical responsibilities are learned and one's vocation is discerned.[169] When access to work is hindered by high levels of unemployment, inadequate systems of training or structural barriers, many young people find the path to their human and professional fulfilment blocked. The need to change jobs several times over the course of life requires that continuous updating and retraining be provided, so that new generations can competently and independently face the risks of an economic environment that is both changing and often unpredictable.[170]

168. This gives rise to a specific public responsibility. The State has the duty to support business activity by fostering conditions favorable to employment, promoting work where it is lacking and defending it in times of crisis, since it is a

169. Cf. United States Conference of Catholic Bishops, *Sons and Daughters of the Light: A Pastoral Plan for Ministry with Young Adults* (November 12, 1996), Washington D.C., 1996, I, 3.

170. Cf. Pontifical Council for Justice and Peace, *Compendium of the Social Doctrine of the Church*, 290.

primary good for families and for society.[171] Particularly in an age of continuous technological transformation, we need a political creativity that will promote "work" and place the family and coming generations at the center; otherwise our economic progress will translate into new forms of insecurity and exclusion.

169. Supporting families and young people in this transition requires choices that make stability feasible. As has been noted above, labor policies need to promote continuity and the quality of employment, countering insecurity as a normal condition of life and encouraging realistic paths for entry into the workforce and for professional growth. Second, measures are needed to ensure a healthy way of living, for without a proper balance between work, leisure and rest, families are weakened and young people struggle to develop a sense of responsibility. Furthermore, it is essential to invest in accessible education and retraining, so that the professional mobility demanded by the digital economy does not become a harsh selection between those who are able to update their skills and those who cannot. Finally, social ties must be supported, with networks and educational communities that accompany life choices and prevent uncertainty from giving rise to loneliness or addictions. If implemented, these technological transformations can be navigated without undermining the capacity to build the future, which is what makes a society prosperous.

---

171. Cf. ibid., 214.

Protecting freedom against dependencies and commercialization

Dependencies and societal control

170. Having reflected on truth and education, work and families, we must now consider the impact of the digital revolution on human freedom, addressing risks to both the mental health of individuals and broader social challenges. The subtler forms of addiction linked to the "digital attention economy" should not be underestimated, since platforms and services are often designed to capture users' time and attention, exploiting their vulnerabilities and weakening their inner freedom. When business models thrive on human weakness, the person is treated as a means rather than as an end; those who design or finance such systems bear a moral responsibility that cannot be ignored. There is an urgent need to promote technologies that strengthen interior freedom by fostering education in digital sobriety and the protection of minors, thus countering models that exploit vulnerability.

171. A further risk, less visible but no less serious, is that of social control made possible by the massive collection of data and use of algorithmic systems. When every action—movements, purchases, relationships and preferences—leaves a trace, a new form of power emerges, namely the power to profile, predict and influence behavior, often without individuals being fully aware of it. If such kinds of data are used to make decisions affecting concrete opportunities—such as

access to credit, employment or essential services—there is a risk of undermining freedom and discriminating against the most vulnerable. Furthermore, control is exercised not only through explicit prohibitions, but also through the architecture of visibility: what is amplified or rendered invisible, what is rewarded or penalized, ultimately shapes opinions and choices, fostering conformity and self-censorship. For this reason, freedom in the digital age is not merely a matter of interiority but also a public concern. It calls for clear rules, transparency, the possibility of recourse and proportionate limits on the use of intrusive technologies, so that technology will remain at the service of the human person and not become a form of control over consciences.

172. At the root of these problems lies a technocratic and post-humanist mentality that tends to regard the human person as an object to be manipulated or a resource to be optimized,[172] removing all safeguards against the unchecked pursuit of profit. What prevails is efficiency, rather than respect for freedom and human dignity. Some post-humanist currents even go so far as to envision "second-class" human beings, subordinate to the interests of elites who consider themselves superior. This troubling prospect becomes all the more serious when combined with technological tools that exponentially increase the capacity for control and selection.

---

172. Cf. Francis, *Message for the Celebration of the 48th World Day of Youth for Peace* (December 8, 2014), 4: *AAS* 107 (2015), 70–71.

Even certain forms of structural indebtedness, which keep entire peoples in conditions of dependence, reflect the same mentality, in new forms, that tolerates relationships of subordination akin to slavery.

Breaking the chains of new forms of slavery

173. This distorted view of the human person is reflected today in various forms of servitude directly linked to the digital economy. Nothing in the world of AI is immaterial or magical. Every seemingly immediate and flawless response is the result of a long chain of mediation, involving vast networks of natural resources, energy infrastructure and, above all, people. A significant part of the digital economy's functioning relies on the silent work of millions of people engaged in essential yet largely unseen activities, such as data labeling, model training and content moderation, often involving disturbing material. In many cases, these workers are young people, predominantly women, working under demanding conditions for minimal wages. Added to this invisible labor is the even harsher work of extracting the resources required for the production of the devices and microprocessors on which AI depends. In some regions of the world, children and adolescents work in dangerous conditions, crushing the materials from which rare earth elements are extracted. The bodies of these people are scarred, injured and worn down so that computational flow may continue uninterruptedly. Furthermore,

criminal networks use online platforms, messaging systems, anonymous payment methods and profiling techniques in order to recruit, control and transport victims of trafficking—very often minors—reducing men and women to "data" to be tracked and "packages" to be moved around within the same digital circuits that support much of the global economy. This reality deeply challenges the moral conscience of our time. It is not enough to invoke efficiency, nor to celebrate the benefits of innovation, if they are built on a chain of exploitation that remains deliberately hidden. If technology promises emancipation, yet produces new forms of global subordination, it stands in contradiction to the fundamental principle of human dignity.

174. The fight against new forms of slavery is a decisive test for the ethical discernment of AI and digital transformation. In continuity with the tradition inaugurated by Leo XIII, the Church renews her firm condemnation of all forms of slavery, trafficking and the commodification of persons. She likewise highlights the urgent need for reflection and action that keep the inalienable dignity of every human being and the common good, as both the focus and goal of society, as well as the guiding criteria for every personal, social and political choice. Without this ethical and humanizing reflection, the growing power of digital systems could lead us toward new atrocities that are no less shameful than those of the past that we now deplore, while we continue to present ourselves as "advanced" and "civilized" societies.

175. Human trafficking must be recognized as a contemporary form of slavery and a grave violation of human dignity. Failing to respond firmly, or tolerating these practices in any way, is in some way to become complicit in today's sins, which are akin to those of the past when slavery was being concealed and justified.[173]

176. In the development of her doctrine, the Church has gradually come to a deeper awareness of the gravity of these issues. It is true that past events cannot be judged anachronistically, as though the moral criteria that matured over time had always been available. Yet neither can we deny or diminish the delay with which both society and the Church came to denounce the scourge of slavery. In antiquity and the Middle Ages many individuals and even ecclesiastical institutions had slaves. Already in the early modern period, the Apostolic See of Rome, responding to requests from Sovereigns, intervened several times in order to regulate and legitimize forms of subjugation, and, in certain cases, the enslavement of "infidels."[174] It was only in the nineteenth

173. Cf. International Theological Commission, *Memory and Reconciliation the Church and the Faults of the Past*, Vatican City 2000, 5.3.

174. As in the Papal Bulls *Sicut Dudum* (January 13, 1435) and *Etsi Suscepti* (January 9, 1442) of Eugenius IV, and in the Papal Bulls *Dum Diversas* (June 18, 1452) and *Romanus Pontifex* (January 8, 1455) of Nicholas V. Political and, at times, even economic needs overcame the demands of the Gospel. The need for evangelization was frequently compromised or at least misunderstood with regard to the needs of worldly powers, thus relativizing the problematic incompatibility of slavery with the Christian conscience.

century that a formal, absolute and universal condemnation of slavery was clearly articulated, notably under Pope Leo XIII.[175] This development offers a clear example of the Church's growth in understanding the perennial truths of Revelation that she safeguards. Although there was not always consistency in practice—given that slavery was long tolerated before being unequivocally condemned—there has been a continuous affirmation throughout history of the dignity of every human being, created in the image of God, even if it took eighteen centuries for its full incompatibility with slavery to be explicitly recognized. This constitutes a wound in Christian memory, one from which we cannot consider ourselves detached.[176] It is impossible not to feel deep sorrow when contemplating the immense suffering and humiliation endured by so many in stark contrast to their immeasurable dignity as persons infinitely loved by the Lord. For this, in the name of the Church, I sincerely ask for pardon.

177. This is why the memory of past complicity and blindness in the face of the injustice of slavery becomes a call to vigilance. What we have learned must be translated into

---

175. Cf. Leo XIII, Encyclical Letter *In Plurimis* (May 5, 1888), *Acta Leonis XIII*, VIII, Rome, 1889, 169–192. Consider that, as late as 1866, the Holy Office distinguished between the immoral and moral aspects of slavery, without fully condemning it: *Instruction of the Holy Office on various doubts of Monsignor Massaia, Vicar Apostolic in the country of the Galla*, April 1866, response to question no. 15.

176. Cf. Saint John Paul II, Bull *Incarnationis Mysterium* (November 29, 1998), 11: *AAS* 91 (1999), 139–141.

discernment and responsibility in the present. If we want to avoid the need to ask for pardon again in the future for having failed to respect the treasure of human dignity that is required by our faith, it falls to us today to denounce, clearly and firmly, trafficking in its many forms and, together with all who are committed to this cause, to support concrete efforts of prevention, protection, liberation and rehabilitation.

178. Even today, colonialism assumes new forms. It no longer dominates only bodies, but appropriates data, transforming personal lives into exploitable information. Entire regions, especially those marked by structural fragility and limited geopolitical relevance, are currently subjected to a new mindset of extraction: that of health data, epidemiological profiles, genetic maps and demographic information. These have become the new "rare earths" of power: vital data which, once aggregated and analyzed, can be used to train predictive models, guide investment strategies, anticipate crises and, above all, determine who and what is deemed to matter. Those who control the health data of entire peoples—often collected under the pretext of aid, research or innovation—possess a structural leverage over the future, for they can shape needs and markets. They can also decide, before others, to whom medicines, investments and protections will be allocated. Here lies one of the most urgent moral challenges of our time: to ensure that shared knowledge becomes a true common good rather than an instrument of dominance. This requires restoring to individuals not only the data that

describes them, but also the ability to decide how it is used, by whom and for whose benefit. Otherwise, the digital age will not be post-colonial, but colonial in another form.

179. New forms of slavery are fueled by economic chains and digital infrastructures. Therefore, action is required on several fronts. First, the supply chains that underpin the technological industry and the digital economy need to become more transparent, so that no competitive advantage is built upon hidden exploitation. Second, companies and investors need to adopt clear criteria for preventive ethical verification (due diligence), placing among their priorities the protection of workers, the fight against forced labor and the assessment of the social impact of data-driven business models. Furthermore, digital platforms must cooperate responsibly with authorities and civil society to prevent communication, payment and profiling tools from becoming channels for the recruitment and control of victims. When such efforts converge, the digital environment can be transformed from a space of exploitation into one of protection, prevention and the promotion of human dignity.

A shared responsibility

180. The various areas just considered—the search for the truth in public life, education in the digital environment, the transformation of work, the fragility of families and new forms of slavery—are not isolated phenomena. Rather, they

reflect a common underlying issue, namely that if technology becomes the ultimate criterion, the human person risks being reduced to data, a cog in a machine or a commodity. If, however, technology is integrated with a wise perspective, it can become an instrument of growth, justice and fraternity.

181. From this perspective, the Social Doctrine of the Church calls for a shared responsibility. It asks that these processes be guided with foresight: by institutions capable of regulating without stifling, and protecting without taking over; by businesses that recognize work and dignity as measures of success; by intermediary organizations and educational communities that rebuild trust and relationships; and by citizens who cultivate responsibility, moderation, discernment and a sense of truth. Only in this way can innovation genuinely serve integral human development, rather than becoming a source of exclusion and dominance. And only in this way can the promise of progress be recognized as authentic, because it is measured against the inviolable dignity of every man and woman.

Chapter Five

The Culture of Power and the Civilization of Love

182. Having considered how AI is transforming certain aspects of life and society, in particular the serious implications for human dignity, we must now turn our attention to the yet more tragic issue of war. Here the question is not merely the efficiency of new tools, but also the risk that technology, detached from ethics and responsibility, will render decisions about life and death more rapid and impersonal, and will present the use of force as an immediate and viable option. In an increasingly interdependent world, peace is not simply one issue among others, but a prerequisite for the universal common good and a test of the moral maturity of peoples, especially of those who bear responsibility for governing.

183. The digital revolution is changing the nature of conflict. Alongside conventional warfare, there are hybrid forms such as cyberattacks, information manipulation, campaigns of

influence and the automation of strategic decisions. AI acts as an accelerating factor in these processes, particularly within a context where many technologies are intrinsically ambivalent. Consequently, what is created for defense can be rapidly repurposed for offense, and the fine line between protection and aggression becomes blurred. While AI can enhance the defense and protection of civilians, it can also lower the threshold for the use of force, shield people from responsibility and foster a culture in which the enemy is reduced to a statistic and the victim to "collateral damage." Faced with these transformations, we must recall the principles of Social Doctrine—the dignity of the person, the common good, the universal destination of goods, subsidiarity, solidarity and justice—for they are criteria for judging whether technologies truly serve humanity or are subjugating it. We should, therefore, consider these principles as guidelines for our decision-making.

184. In this chapter, therefore, I will compare two opposing approaches, which I have already evoked through biblical imagery in the Introduction. On the one hand, there is the temptation of constructing the Tower of Babel, relying on power and pride. On the other hand, patience is required in order to rebuild Jerusalem "piece by piece," as in the time of Nehemiah, by safeguarding humanity and the common good.

185. If we examine global dynamics, we can recognize more clearly the spread of a culture of power characterized by polarization and violence. The modern Babel can be seen

not only in the globalized technocratic paradigm, but also in the remote clash between opposing imperialisms, between powers that wish to preserve their supremacy, and those that aspire to seize that supremacy, resulting in a multiplicity of local conflicts. Moreover, there seems to be no limit to the race—driven by a dehumanizing ambition—to develop evermore powerful technologies or to secure control over them. Yet, despite this downward spiral, we can also glimpse a great part of humanity that is striving to remain human and working to build the holy city of coexistence and peace. All too often, we are unwitting builders and clumsy architects of this city, capable of generous gestures but lacking an overall vision. This building project is slower, less visible and less spectacular, and awaits a better understanding and greater coordination so that it may become the conscious and clear responsibility of every community, from families to States, and the relations between Nations. It is this prospect of commitment, this construction site of hope, that we call the "civilization of love."

The civilization of love in the digital age

186. When Saint Paul VI coined the phrase "the civilization of love,"[177] the world was in the midst of the Cold War,

177. Cf. Saint Paul VI, *Regina Caeli* (May 17, 1970): *Insegnamenti di Paolo VI*, vol. VIII, 506.

an arms race and severe economic instability. In that context, the Church proposed an alternative path to that of ideological opposition between systems, and envisioned a social order in which justice and charity are intertwined and love becomes the guiding principle of economic, political and cultural life. Today, we must resolutely recover this vision, for the civilization of love is no naïve utopia, but a demanding project, which consists in translating charity into structures of justice, giving institutional form to fraternity and regarding others—whether individuals or peoples—as allies necessary for building the common good. As the Encyclical Letter *Fratelli Tutti* reminded us, only this social love is capable of becoming a culture and a norm, and thereby of bringing about a stable international order, transforming mere armed coexistence into a community with a shared future.[178]

187. This insight proves even more fundamental in the current context of digital transformation. Digital networks, the globalized economy and the development of AI create increasingly tighter bonds, linking—in real time—decisions made in one place to the effects they produce elsewhere. In this sense, the words of the Second Vatican Council on the growing interdependence between peoples remain timely, for the common good is taking on an increasingly universal

178. Cf. Francis, Encyclical Letter *Fratelli Tutti* (October 3, 2020), 183: *AAS* 112 (2020), 1033–1034.

dimension, with rights and duties concerning the entire human family.[179] The project for a civilization of love, therefore, must undertake the task of transforming this imposed interdependence into a willed and chosen solidarity. This is the guiding principle for technological processes: it is not enough for artificial intelligence to make us more efficient or connected; it must also serve to build a universal human family, with shared rights and duties, where digital proximity becomes a real opportunity for encounter and mutual care.

The culture of power

188. In our time, a culture of power is taking hold, in which the availability of resources and the ability to dominate tend to dictate the agenda and criteria for decision-making. In this way, the common good of humanity is relegated to the background and the concrete tragedy of peoples at war is reduced to a secondary consideration in relation to strategic interests. This culture of power infiltrates society, changes relationships and behaviors, and grows by normalizing war, pursuing ever-greater military power, taking advantage of the crisis of multilateralism and fueling a false realism that insists that there is no alternative.

179. Cf. Second Vatican Ecumenical Council, Pastoral Constitution *Gaudium et Spes*, 26: *AAS* 58 (1966), 1046–1047.

The normalization of war

189. In 1965, the words of Saint Paul VI resounded powerfully at the UN General Assembly: "Never again war, never again war!"[180] We must acknowledge that, despite the desires and declarations for peace, the past sixty years have been marked by conflicts of astonishing brutality, often affecting civilian populations on a massive scale, leading to the death of innocent victims, mass displacement, social destabilization and long-lasting wounds. Nevertheless, in public discourse, there was a widespread conviction that war should remain a last resort, subject to strict ethical and legal limits, and always oriented toward a political vision of peace. Following developments in the immediate post-First World War period, a turning point occurred after the Second World War: peace was made the focus of the international order, as attested in particular by the United Nations Charter, with the intention to "save succeeding generations from the scourge of war."[181] Likewise, many national constitutions restricted the use of force to extreme and strictly limited circumstances. Even during the Cold War, despite the existence of serious conflicts, there remained the awareness that a new world war had to be avoided at all costs.

---

180. Saint Paul VI, *Address to the 20th General Assembly of the United Nations* (October 4, 1965): *AAS* 57 (1965), 881.

181. United Nations, *United Nations Charter*, San Francisco (June 26, 1945), Preamble.

190. Today, however, we are witnessing a real paradigm shift in public discourse and in decisions regarding rearmament, with a troubling revival of war as an instrument of international politics, while the very ethical principles that had previously limited its use are being eroded. Regional conflicts that drag on over time, escalating tensions and reciprocal threats are becoming almost commonplace, and forms of conflict driven by the desire for territorial expansion that were thought to be overcome are re-emerging. Public opinion is gradually being shaped and conditioned by polarizing media narratives, which are often amplified by algorithms that prioritize conflict and confrontation.

191. We are also witnessing a disconcerting loss of historical memory, as first-hand accounts of the Holocaust and the two World Wars are disappearing. This leads to a selective or distorted rewriting of the past, in a context where fake news and the manipulation of narratives obscure the lessons that have been learned. Without a living memory of the horrors of war, political decisions risk being made on the basis of power alone, without any consideration for the long-term consequences.

192. To all of this, the media and digital dimensions are adding new and decisive elements. Communication networks, fragmented information environments and algorithms that reward conflict can magnify polarization and resentment, increase propaganda and make shared discernment more difficult. Thus, war is not only fought, but also culturally

conditioned through simplistic narratives, a friend-or-foe mentality, disinformation and fear. When historical memory fades and the ethical principles that protect civilians and the most vulnerable are weakened, it becomes easier to justify violence as necessary, inevitable or even "sanitized." It is in this context that humanity is slipping into a violent culture of power, where peace no longer appears as a responsibility to be taken on, but as a fragile interval between conflicts. Today, more than ever, without prejudice to the right to self-defense in the strictest sense, it is important to reaffirm that the "just war" theory, which has all too often been used to justify any kind of war, is now outdated.[182] Humanity possesses far more effective and capable tools for promoting human life and resolving conflicts, such as dialogue, diplomacy and forgiveness. The use of force, violence and weapons reflects a relational poverty that always has disastrous consequences for civilian populations.

---

182. Cf. Francis, Encyclical Letter *Fratelli Tutti* (October 3, 2020), 258: *AAS* 112 (2020), 1061: "In recent decades, every single war has been ostensibly 'justified.' The Catechism of the Catholic Church speaks of the possibility of legitimate defense by means of military force, which involves demonstrating that certain 'rigorous conditions of moral legitimacy' have been met. Yet it is easy to fall into an overly broad interpretation of this potential right. In this way, some would also wrongly justify even 'preventive' attacks or acts of war that can hardly avoid entailing 'evils and disorders graver than the evil to be eliminated.'"

Force without limits

193. The growth of the military-industrial complex has become a defining feature of the current political landscape and has become a key sector in the economy of various countries. The close link between economic interests, the military apparatus and political decisions produces an "armed nation," in which war appears as a natural extension of politics, and the arms market becomes an autonomous driving force behind military decisions. Nor can we ignore the enormous economic interests behind war. The armaments industry, and countries that supply weapons, profit from a market that thrives precisely on conflicts. In this sense, there are also financial interests that contribute to fueling tensions in various regions of the world.

194. Military arsenals are receiving renewed attention. In the past, recognition of the threat posed by weapons capable of destroying all of humanity had promoted paths toward *détente* and disarmament negotiations. Unfortunately, this approach has been left behind, and the evolution of nuclear arsenals—including the prospect of its "tactical" use—makes the use of such weapons seem less improbable. In this context, the Treaty on the Prohibition of Nuclear Weapons, which came into force in 2021 with the support of over seventy countries, is an important step. However, it risks remaining largely symbolic since the major nuclear powers have not agreed to it. This has led to the widespread yet erroneous

belief that nuclear deterrence is an indispensable prerequisite for security. This has also contributed to a new arms race, which is hard to control and accompanied by the gradual dismantling of nuclear reduction agreements, as well as the development of "miniaturized" weapons, that make their use seem like a more viable option.

195. The same logic applies to conventional warfare. Military force, weak diplomatic initiatives and the complexity of the interests at stake contribute to conflicts that tend to become protracted, with extremely high human and environmental costs. It is much easier to start a war than to stop it, and yet, discussion on conflict prevention remains tragically marginal.

196. The situation is further destabilized by the presence of new armed operatives, such as jihadist groups, private militias and criminal networks that mark the end of the State's monopoly on the use of force. Often these groups intertwine vague ideological motivations with concrete economic interests, transforming war into a "way of life" for entire generations of young people and children. Here, the objective is no longer a definitive victory, but the perpetuation of conflict as a source of power and income.

Weapons and artificial intelligence

197. The above-mentioned scenario is linked to the unceasing development of weapons systems, particularly

those involving AI. The Holy See has recently observed that the growing ease with which autonomous weapons systems can be deployed makes war more "feasible" and less subject to human control. This violates the principle that armed force should be used only as a last resort in cases of legitimate self-defense.[183] For this reason, the development and use of AI in warfare must be subject to the most rigorous ethical constraints, to guarantee respect for human dignity and the sanctity of life and to avoid a race to develop such arms.[184]

198. Sometimes there is talk of "artificial moral agents," as if machines were able to distinguish between right and wrong with greater consistency than a human being. Yet moral judgment cannot be reduced to calculation, for it involves conscience, personal responsibility and the recognition of the other as a person. Therefore, it is not permissible to entrust lethal or otherwise irreversible decisions to artificial systems. No algorithm can make war morally acceptable. AI does not remove the intrinsic inhumanity of conflict; indeed it can only bring about conflict more quickly and render it more impersonal, lowering the threshold for resorting to violence, transforming defense into threat prediction and thus reducing victims to data. In this way, it will accustom us to the idea that

---

183. Cf. Dicastery for the Doctrine of the Faith and Dicastery for Culture and Education, Note *Antiqua et Nova* (January 14, 2025), 99: *AAS* 117 (2025), 202–203.

184. Cf. ibid., 103: *AAS* 117 (2025), 204.

violence is inevitable and needs only to be optimized. This does not diminish the importance of instilling, as far as possible, values and sound judgment into the artificial systems we build, so that they can contribute to a moral ecosystem in which humans are better able to listen to their own consciences, as well as allowing AI models to establish appropriate boundaries.

199. It is not enough to invoke a generic type of ethics. Concrete criteria for discernment must be established. The first such criterion concerns personal responsibility. When a decision to strike becomes automated or opaque, the risk of abdicating responsibility increases. For this reason, the chain of responsibility must be identifiable and verifiable; those who design, train, authorize and employ technology must be held accountable for their decisions. The second criterion pertains to the moral timeframe for making judgments. While AI tends to expedite the decision making processes, speed and efficiency should never be the supreme motivating force for the irreversible decisions made in the context of war. The third criterion is the identification and protection of civilians. Any technology that facilitates attacks without seeing the face of human beings lowers the moral threshold of conflict. Target selection and the use of force must not confuse combatants and non-combatants, nor ignore the impact on defenseless populations.

200. These criteria give rise to certain non-negotiable requirements. First, all systems used in a war setting must

guarantee the possibility of retracing and reconstructing decision-making processes, so that accountability and blame are not collapsed into "the machine." Second, the decision to use lethal force cannot be delegated to opaque or automated processes, but must remain under effective, self-aware and responsible human control. Finally, it is imperative to establish a shared framework—also at the international level—in order to curb the technological arms race and ensure robust protection for civilians and the infrastructures necessary for their survival.

The crisis of multilateralism

201. The culture of power also stems from the crisis of the multilateral system. The institutions established to safeguard the concept of a common future for all peoples and a global common good appear to have been weakened. This is due not only to structural limitations, but also to a frequent lack of shared will to support and reform them, or to recognize their moral authority. Instead of making progress, we are regressing from the significant turning point of the twentieth century. After 1989, the collapse of communist regimes in Europe was followed by a predominantly economic globalization, which lacked an adequate political framework capable of sustaining dialogue and peace. An almost blind faith was placed in the ability of the markets to generate prosperity, democracy and stability. In reality, rather than automatically generating unity

and peace, globalization has provoked fundamentalist, identity-based and nationalistic reactions. The result is a far cry from genuine multilateralism; instead, what has appeared is a disorderly and conflict-ridden multipolarism with a prevailing sense of mistrust.

202. What has also re-emerged is the temptation to forge a collective identity in opposition to an enemy, fueled by narratives in which each party portrays itself as a victim entitled to retribution. The reduction of complex issues into simplistic categories—"me first," "friend or foe," "us or them"—facilitates decisions that are often irresponsible and undermine mutual trust among nations. The force of international law is thus replaced by the claim that "might makes right." Consequently, tribunals that are competent for settling disputes between States or dealing with war crimes are often weakened or bypassed, with devastating ramifications for political culture and social cohesion.[185]

203. In this context, peacebuilding has been relegated to a secondary role. Cooperation for development, disarmament, conflict prevention and the establishment of mutual trust are neglected in the name of power politics. The achievements of humanitarian law are also being compromised. Indeed, the principle of proportionality in responding to aggression, the

---

185. Cf. Leo XIV, *Address to the Participants in the Plenary Session of the "Reunion of Aid Agencies for the Oriental Churches (ROACO)"* (June 26, 2025): *AAS* 117 (2025), 847–849.

protection of access to water, food and essential goods, and respect for the lives of civilians, especially children, come to be regarded as naïve relics of the past.

A supposed political realism

204. We live at a time of significant spiritual and cultural blindness. A false pragmatism urges us to sever the roots of our history, as if it were possible to inaugurate a kind of "new creation" detached from the past. Even those who cite important moral principles can fall into this historical nihilism, mistakenly believing that the atrocities of the twentieth century can never happen again. Yet, in reality, the same dynamics are re-emerging under new guises. The mentality of armed equilibrium and deterrence appears to be reasserting itself. Today, however, in contrast to the two-sided dynamic of the Cold War, the proliferation of operatives and battlefields makes this mentality increasingly fragile. Escalating conflicts lead to asymmetric and "hybrid" wars, fought not only on the battleground but also on the economic, financial and cyber fronts, where disinformation and campaigns that feed people's fears are used to manipulate public opinion. In many countries, including those in the Global South, increased military spending is presented as the only response to an uncertain future or perceived threats. Meanwhile, the real cost falls on the poorest, who see resources for healthcare, education and social services being reduced.

205. At the core of these issues is a false realism, based not only on the prevailing mentality of force, but on the cultural and anthropological belief that war is an inevitable part of human nature. It is said that things have always been this way, except for occasional pauses, and that it will always be so! As a result, the concern is no longer the search for peace—which has been lost as a point of reference on the international stage—but rather how and when to take military action. This same argument maintains that it would be irresponsible not to prepare for conflict. I would argue, however, that what is truly irresponsible is *Realpolitik*, the form of political "realism" that sows in consciences and in society an attitude of resignation to the inevitability of war, and dismisses peace and dialogue as utopian or irrational positions that ignore the risks at stake. In fact, peace is neither a naïve hope nor merely the absence of war; instead, it is always possible as the fruit of justice and charity.

206. In such a climate, nihilism and pragmatism become intertwined and end up normalizing grave errors. Religious extremism and identity-based fanaticism ally themselves with irrational economic policies, while politics often turns to misinformation and ridiculing opponents, and systematically cultivating fears and resentments. Thus, diversity is increasingly perceived as a threat, which fuels a desire for possession, a will to dominate, hegemonic ambitions, abuses of power and a fear of those who are different, thereby creating an

environment in which new conflicts can develop almost imperceptibly.[186]

207. This, then, is the fertile ground for new wars that are perhaps even more dangerous than those of the past, since they tend to disregard all ethical limits. What was once considered unacceptable can now be carried out almost without hesitation, while the international response is increasingly influenced more by the interests of individual Governments than by the objective gravity of situations. Decisions now seem to be driven almost exclusively by economic calculations, justified through media distortions, manufactured enthusiasm and "dreams" that inevitably shatter, generating frustration and further violence. When people come to believe that nothing is genuinely true and that principles are hollow words, then the fuse in their hearts is lit for new eruptions of intolerance and aggression.

208. In these situations, the issue of concrete safeguards to prevent future violence remains an open question. When a culture normalizes and justifies conflict, a dangerous pathway opens up, in that what seems unthinkable today may become acceptable tomorrow in the name of utility or security. In countries marked by serious social tensions, we cannot rule out the possibility that some leaders may consider armed conflict as an

---

186. Cf. Francis, *Message for the 53rd World Day of Peace* (December 8, 2019): *AAS* 112 (2020), 54–61.

effective way of diverting attention from domestic problems and a cynical tool for managing difficulties.

209. A particular responsibility rests on the shoulders of those who work in the field of research. All the key players in this field—scientists, business owners, investors, academic authorities, politicians and others—must work with a transparent and responsible mindset, while maintaining an acute awareness of the broader context of the technological advancements they help to cultivate, including those related to AI. When people limit themselves to looking only at their own sector, they may deceive themselves into believing they are performing actions that are morally neutral and avoid questions about the ultimate ends that guide certain experiments. In this way, they risk cooperating—perhaps unknowingly—with questionable projects that fuel new forms of violence, manipulation and dominance.

Building the civilization of love

210. The construction of a world in a state of perpetual conflict is an evil and must be named for what it is. This way of portraying our current situation may seem bleak or pessimistic, yet I consider it necessary to do so. The Christian perspective, however, is not limited to denouncing evil. We view history in the light of the crucified and risen Lord, to whom the Father has given "all authority in heaven and on earth" (Mt 28:18). We do not consider the present as a

predetermined fate, but an opportunity for personal and collective conversion. Moreover, we believe in the power of the Kingdom, which grows from the tiny size of a mustard seed, which, once sown, sprouts and grows (cf. Mk 4:26–32). While the tumult of confusion is all around us, goodness grows silently from the earth. In the words of the prophet Isaiah: "Behold, I am doing a new thing; now it springs forth, do you not perceive it?" (Is 43:19).

211. A closer analysis of history confirms this. Even in the darkest nights, the Lord raises up men and women who refuse to give up, who persevere in doing good, who protect the vulnerable and open pathways to reconciliation. The memory of the saints, righteous people and the oft-forgotten peacemakers, show us that grace does not magically eliminate conflict, but instead it inspires active resistance to evil and an astonishing creativity in doing good. Christians see the darkness and acknowledge it for what it is, yet they do not merely gaze upon it passively, for they know the light and understand that the darkness has not overcome it and cannot defeat it (cf. Jn 1:5). For this reason, even when suffering seems to have the last word, Christians serve the good and are sustained by a theological hope that gives reality both meaning and direction.

We can all do our part

212. At this point, however, a subtle temptation may emerge, namely the thought that the problems are too big

and we are too small, and that our choices, therefore, cannot make a difference. This is a polite form of resignation, often disguised as realism. Certainly, not everyone has the same power to make a difference. There are those who govern, make investment decisions, lead institutions, conduct research, educate, produce or provide information, and then there are those who only seem to live their daily lives. Yet, no one is without responsibility. We all have our own areas for action, and it is precisely there—and nowhere else—that we must choose whether to fuel the mentality of force (even if only through indifference, cynicism, lies or hatred), or to preserve the mindset of peace (with truth, moderation, closeness and care).

213. The twentieth-century Catholic author J. R. R. Tolkien, in the words of a protagonist in one of his novels, described our responsibility in this way: "It is not our part to master all the tides of the world, but to do what is in us for the succour of those years wherein we are set, uprooting the evil in the fields that we know, so that those who live after may have clean earth to till."[187] The civilization of love will not arise from a single or spectacular gesture, but from the sum total of small and steadfast acts of fidelity that serve as a bulwark against dehumanization. For this reason, it is worthwhile pausing to reflect on some aspects of how we,

187. J. R. R. Tolkien, *The Lord of the Rings. The Return of the King*, Part III, Book Five, Chapter IX, New York 1965, 190.

each in our own way, can cooperate in building the civilization of love. Without presuming to exhaust this theme, I would like to propose five paths toward daily and public responsibility: the need to disarm words, building peace through justice, adopting the perspective of victims, cultivating a healthy realism and reviving dialogue and multilateralism.

The need to disarm words

214. The first contribution we can make toward a more humane civilization is to be mindful of our words. "Let us disarm words and we will help to disarm the world."[188] Words have enormous power, something we experience in our daily interactions; for example, spoken words can change our mood for better or for worse. "Peace begins with each one of us: in the way we look at others, listen to others and speak about others. In this sense, the way we communicate is of fundamental importance: we must say 'no' to the war of words and images, we must reject the paradigm of war."[189] We must all, therefore, examine our conscience regarding the words we use, the prejudices we have and the explicit or implicit aggression that lies within them. We have a real opportunity to contribute to the common good each time we speak the truth,

188. Leo XIV, *Address to Representatives of the Media* (May 12, 2025): *AAS* 117 (2025), 682.

189. Ibid.

offer wise advice, support those in need of comfort, denounce injustice and give a voice to the voiceless.

Building peace through justice

215. All of us, at every level, can contribute to building the foundation of peace, which is justice. We do not merely seek any kind of peace—such as an absence of conflict at any cost—but instead, the true peace born of justice. "There exists a very close connection between the justice of the individual and the peace of everyone."[190] Commenting on the psalm verse "justice and peace have embraced" (Ps 84:11), Saint Augustine wrote: "There is no one who shuns the desire for peace, yet not everyone is willing to practice justice . . . But perform the works of justice, keeping in mind that justice and peace have embraced; they are not at odds with one another. Why do you set yourself against justice? Here, for example, is justice telling you not to steal, but you pay no heed; not to commit adultery, and you turn a deaf ear; not to do to others what you would not want done to yourself; not to say about your neighbor the things you would not want said about yourself . . . Do you therefore wish to attain peace? Then practice justice!"[191] Let us never grow weary of seeking justice!

---

190. Saint John Paul II, *Message for the 31st World Day of Peace* (January 1, 1998), 1: *AAS* 90 (1988), 147.

191. Saint Augustine, *Enarrationes in Psalmos*, 84, 12: *CCSL* 39, Turnhout 1956, 1172–1173.

Adopting the perspective of victims

216. There are times when, in order to remain human, we must set aside our reservations and take a stand. In some conflicts, it is unjust to remain neutral, nor is it enough merely to claim that we are not complicit.[192] When we witness the bombing of civilians, attacks on hospitals, schools or vital infrastructure, and violence that affects children, we are confronted with scandals that wound humanity itself. For this reason, we cannot limit ourselves to the level of abstract analysis. Pope Francis encouraged us to "touch the wounded flesh"[193] of those who suffer, look at their faces, listen to their stories and acknowledge their wounds. Painful events require both history and memory, the former to recount the facts, the latter to bear witness to lived experiences.

217. Giving space to the perspectives and voices of victims through communication and education helps us to become aware of the abyss of evil inherent in war, and generally in all forms of violence. It helps us to reject the normalization of conflict; not to turn away when human dignity is violated; and to restore to victims the dignity of being recognized and heard.[194] Paying attention to these voices

192. Cf. Francis, Encyclical Letter *Dilexit Nos* (October 24, 2024), 22: *AAS* 116 (2024), 1375–1376.

193. Francis, Encyclical Letter *Fratelli Tutti* (October 3, 2020), 115: *AAS* 112 (2020), 1008–1009.

194. Cf. ibid., 261: *AAS* 112 (2020), 1062.

strengthens the conviction that, apart from violent minorities, humanity does not desire war. In a particular way, the Church can be a place of living memory for victims. As Saint Paul VI recalled, the Church feels she must make her own both the voice of those who died in past wars and the voice of the living who still bear wounds today, so that their cries may become an appeal for peace and harmony and not a prelude to new conflicts.[195]

Cultivating a healthy realism

218. We are in need of a healthy realism that avoids both political idealism and cynicism. There is a kind of idealism that, in order to preserve its own worldview, tends to choose facts selectively, distorting and renaming them. Its proponents eventually, inhabit a reality constructed to fit their own convictions. Conversely, there is also a debased form of realism that confuses observation with resignation, arguing that since force prevails, it will always prevail. Authentic realism does not give up on changing the world; indeed, it starts by clearly identifying interests, fears, constraints and power dynamics, precisely in order to determine what can be achieved, and the measures needed to achieve it. It does not reduce politics to morality; neither does it surrender to

195. Cf. Saint Paul VI, *Address to the 20th General Assembly of the United Nations* (October 4, 1965): *AAS* 57 (1965), 878–879.

violence. Instead, it seeks viable paths for making peace more than a mere word, through credible institutions, verifiable guarantees, patient negotiations, conflict prevention and the protection of civilians.

Reviving dialogue

219. In order to build the civilization of love, we must engage in dialogue, for this is the primary means of coexistence between people and nations, and it is the alternative to open conflict. On the eve of the Second World War, Pius XII affirmed that nothing is lost with peace, whereas with war everything can be lost. He insisted that people must return to speaking with one another, because a sincere and persevering dialogue always opens up the possibility of an honorable solution.[196]

220. Indeed, dialogue is an ordinary part of human life and does not only concern relations between States. It involves acquiring an attitude that seeks to forge bonds of fraternity built on listening, an open demeanor, making time for each other and even wasting time together. For if we experience authentic encounters with others, with those who are different, strangers and migrants, it becomes much more difficult even to imagine war.

---

196. Cf. Pius XII, Radio Message *A Grave Hour* (August 24, 1939): *AAS* 31 (1939), 334.

221. At the political level, there is an urgent need to shift from the "culture of power" to a genuine "culture of negotiation," in which dialogue and diplomacy become the standard means of resolving conflicts. Giorgio La Pira expressed the hope that "the method of war be replaced by the method of peace: the method of negotiation, of encounter, of convergence, that is, the authentically human method!"[197] The awareness that all peoples share a common future demands that the "culture of negotiation" become an increasingly shared political and cultural commitment, capable of gradually leading humanity away from the cycle of violence.

222. To those who have the honor and responsibility of governing, I would like to repeat the words that I spoke at the start of my Pontificate: "The peoples of our world desire peace, and to their leaders I appeal with all my heart: Let us meet, let us talk, let us negotiate! War is never inevitable. Weapons can and must be silenced, for they do not resolve problems but only increase them. Those who make history are the peacemakers, not those who sow seeds of suffering. Our neighbors are not first our enemies, but our fellow human beings; not criminals to be hated, but other men and women with whom we can speak. Let us reject the Manichean notions so typical of that mindset of violence

197. Giorgio La Pira, *Riflessioni sul Concilio*. Address of Professor Giorgio La Pira, Mayor of Florence, to the "Guides de France" (Rome, September 4, 1962), Florence 1962, 6.

that divides the world into those who are good and those who are evil."[198]

223. In rejecting the mindset of violence, interreligious dialogue plays a decisive role, because at the heart of the great spiritual paths lies a message of peace.[199] Whereas those who use the name of God to legitimize terrorism, violence or war betray his true nature, for to fight in the name of religion means attacking religion itself.[200] The "spirit of Assisi," evoked by Saint John Paul II and carried forward by Pope Francis—for example, through his dialogue with the Grand Imam of Al-Azhar—shows that believers can draw upon the most authentic sources of their particular spiritual traditions, where there is no room for "sanctified hatred."

The necessity of diplomacy and multilateralism

224. In international relations, dialogue is an irreplaceable diplomatic tool for preventing conflicts and rebuilding bonds of trust. Faced with the impulsive broadcasts, aggressive rhetoric and power politics that characterize our time, "the

---

198. *Address to Participants in the Jubilee of Oriental Churches* (May 14, 2025): *AAS* 117 (2025), 686.

199. Cf. Francis, Encyclical Letter *Fratelli Tutti* (October 3, 2020), 271: *AAS* 112 (2020), 1066.

200. Cf. Francis, *Appeal for Peace at Assisi for the World Day of Prayer for Peace "Thirst for Peace: Faiths and Cultures in Dialogue"* (September 20, 2016): *AAS* 108 (2016), 1124.

vocation of diplomacy is to foster dialogue with all parties, including those interlocutors considered less 'convenient' or not considered legitimized to negotiate."[201] Therefore, every ounce of humility and patience should be employed in order to nurture even the faintest signs of goodwill among parties in conflict, so as to advance the process of peace.

225. Cyberspace too has become a battleground. Cyber-attacks, data manipulation and campaigns of influence, orchestrated with the help of AI, can destabilize entire countries even before open armed conflict erupts. Moreover, in this area, the attribution of responsibility is often uncertain. When it is unclear who carried out an attack, the risk of disproportionate reaction, miscalculation and escalation increases. For this reason, diplomacy must be capable of operating effectively in this new environment, negotiating shared regulations on the use of digital technologies, in order to protect civilians and the most vulnerable from "invisible" yet real forms of violence.

226. International organizations, particularly the United Nations, are essential instruments for promoting a civilization of love, for they can foster dialogue among nations and promote the peaceful resolution of conflicts, the integral development of peoples, the protection of the most vulnerable, disarmament and the care of creation. Through such efforts, the international community can work to reduce inequalities,

---

201. Francis, *Address to Members of the Diplomatic Corps accredited to the Holy See* (January 9, 2025): *AAS* 117 (2025), 110.

defend the rights of refugees and minorities, reallocate resources from military spending to human development and protect our common home. The Holy See supports and accompanies these endeavors, while also recognizing that the current weaknesses of the UN and the international political system reveal the need for profound reforms. This is not simply a question of technical adjustments, for the crisis of convictions and values that also concerns the ethical foundations of nations makes it more difficult to direct multilateralism toward the true common good.[202]

227. In the international context, the Holy See's diplomacy adopts the Gospel's principle of mercy as a concrete criterion for political action. This is one of the ways in which the Holy See places itself at the service of humanity, thereby appealing to consciences in the name of charity and truth, defending the dignity of every person and speaking up on behalf of the poor, migrants and victims of war. In this way, papal diplomacy expresses the catholicity of the Church and contributes to the building of a civilization of love, where even new technologies can be oriented toward the common good.

Praying and hoping

228. These avenues for exercising responsibility are sustained by prayer, and in turn nourish prayer. Indeed, for each

202. Cf. Francis, *Address to the Participants in the 38th Conference of the FAO* (June 20, 2013): *AAS* 105 (2013), 616–617.

of us, peace primarily comes "from God, God who loves us all, unconditionally."[203] It is a gift given by Jesus to his disciples on the day of Easter: "Peace be with you! It is the peace of the risen Christ. A peace that is unarmed and disarming, humble and persevering."[204] With these words, I greeted the Church and the world on the day of my election to the See of Peter. I wish to repeat them now, and to invite everyone to pray for this gift. Let us never tire of praying for peace and of committing ourselves to achieving it in our relationships and in society.

---

203. Leo XIV, *First "Urbi et Orbi" Blessing* (May 8, 2025): *AAS* 117 (2025), 660.

204. Ibid.

Conclusion

229. "Let each builder choose with care how to build" (1 Cor 3:10). With these words, Saint Paul encouraged the Christians of Corinth to preserve unity. Dear brothers and sisters, we have reflected on the world we are building, and we asked ourselves what it means to safeguard the human person in the era of artificial intelligence. At the end of this reflection, I would like to propose a sober yet demanding program of Christian life with which we can navigate this epochal change in the light of the Gospel. This avenue emerges through contemplating God's plan, living ecclesial unity by partaking of the Eucharist, building a world centered on the common good and praying in union with the Blessed Virgin Mary.

The Word became flesh

230. Our world is filled with attempts to seize control of markets and spheres of influence, often shrouded in reassuring rhetoric and seductive ideologies. Yet our hearts yearn for

an approach that is wise and benevolent, akin to that which Mary praises in her *Magnificat*, when she proclaims that God's mercy extends in every generation to those who fear him.[205] This plan of mercy continues to unfold throughout history today, even amid the rapid and unsettling changes brought by algorithms and global networks, and it becomes a compass in the digital era for living our lives according to the Gospel.

231. At the heart of everything is the mystery of the Incarnation, the Word who became flesh and dwelt among us. The flesh of the Son, poor and vulnerable, evokes the flesh of so many brothers and sisters stripped of their dignity and reduced to silence.[206] Through the Lord's closeness, the gift of peace enters into the world in a paradoxical way. It does so through the power to become children of God, and is awakened when we allow ourselves to be moved by the tears of the little ones, the fragility of the elderly, the silence of victims and the struggle of those who fight against the evil they do not wish to commit.[207] In this wounded yet beloved flesh, the Father shows us the true humanity of a life fulfilled through

205. Cf. Leo XIV, *Homily at First Vespers on the Solemnity of Mary the Most Holy Mother of God* (December 31, 2025): *L'Osservatore Romano*, January 2, 2026, 1–2.

206. Cf. Leo XIV, *Homily of the Mass during the Day* (December 25, 2025): *L'Osservatore Romano*, December 27, 2025, 3.

207. Cf. ibid.

openness and communion, which leads us to desire that his will be done on earth as it is in heaven.[208]

232. In the promises of transhumanism and some post-humanist currents of thought, which seek an enhanced and almost disembodied humanity, we recognize a yearning that is of concern to us, namely the need for a fuller life, less exposed to limitations and suffering. Yet the Incarnation opens a different pathway. On the one hand, old and new ideologies alike urge humanity to overcome limitations through technology, and to rise above others by asserting dominance. Contrary to this, the mystery of the Son of God entering into our human condition promises something quite different. The living God descends into our history in order to free us from all forms of slavery.[209] He takes upon himself our weakness and transforms it into a setting for salvation. There is no moment or human situation that is not worthy of God. "According to the teaching of our faith, we have and adore, in our mysteries, a God who is born in a manger, a God who lives and travels in Judea, a God who dies on the cross, a dead God who lies in the tomb."[210] The future of humanity,

---

208. Cf. Leo XIV, *Angelus on the Solemnity of the Epiphany* (January 6, 2026): *L'Osservatore Romano*, January 7, 2026, 3.

209. Cf. Leo XIV, *Homily of the Mass during the Night* (December 24, 2025): *L'Osservatore Romano*, December 27, 2025, 2.

210. P. de Bérulle, *Discours de l'état et des grandeurs de Jésus*, *Discours IV*, *Unité de Dieu en l'incarnation*: *Œuvres complètes*, Paris 1856, col. 218.

therefore, finds its standard in the ability to welcome this divine way of drawing near, of sharing the burden of the world, of transforming relationships from within. "O wonder . . . man is God and this God-Man passes through all those stages, endures all those states and ennobles them, sanctifies them, deifies them in himself!"[211] What saves humanity is the divine love that descends into the most fragile point of our history and renews it from within.

233. For this reason, as a believer among believers, I invite everyone to contemplate, in the face of the Son of God, the *grandeur of humanity* that shines a light also on the era of AI. In Christ, we are called to cooperate in the work of creation, rather than be disinterested observers of technological processes that limit our freedom and responsibility.[212] The dignity inscribed in each of us by the Holy Spirit can also be seen in our capacity to reflect critically, choose and love freely, and form authentic relationships. No computational system, however sophisticated, can create a heart that gives itself, or a conscience that discerns good from evil. Even when machines excel in efficiency, a human face that asks to be gazed upon remains the center of our history. This human face is the fullness toward which history is moving. It is the mystery of

211. Ibid.

212. Cf. Leo XIV, *Address to the Conference "Artificial Intelligence and Care of Our Common Home"* (December 5, 2025): *L'Osservatore Romano*, December 5, 2025, 2.

"recapitulation": the certainty that the Father has decreed to bring all things, those in heaven and those on earth, back to Christ, the one Head (cf. Eph 1:10). In this plan, nothing will be lost that is authentically human. Indeed, everything will be purified and reunited in the One, who gathers every fragment of life, every tear and every authentically human achievement, rescuing them from nothingness and delivering them, redeemed, to the Father.

One body in Christ

234. The spirituality that we need is a Eucharistic spirituality, that is, a spirituality of ecclesial unity in love. The Incarnation and the Paschal Mystery reveal God entering into our human condition and transforming it through the gift of himself. This gift remains present and active in the Eucharist, in which the Lord gives himself and gathers the Church together, so that his offering becomes the principle of unity and source of new life. It is from this communion that Christian solidarity also arises, since "union with Christ is also union with all those to whom he gives himself."[213] As Saint Augustine explained to the new Christians of his local Church, the bread and wine on the altar are the sacrament of

213. Benedict XVI, Encyclical Letter *Deus Caritas Est* (December 25, 2005), 14: *AAS* 98 (2006), 228.

the unity of the faithful in Christ: "What is seen is a mere physical likeness; what is grasped bears spiritual fruit. So now, if you want to understand the body of Christ, listen to the Apostle Paul speaking to the faithful: *together you are the body of Christ* (1 Cor 12:27). If you are the body and members of Christ, then it is your sacrament that is placed on the table of the Lord; it is your sacrament that you receive. You respond 'Amen,' and by responding in this way you assent to it. For you hear the words, 'the Body of Christ' and respond 'Amen.' Be then a member of the Body of Christ that your Amen may be true!"[214]

235. The "Amen" that we say in the liturgy, the Body we eat and the Blood we drink shape our entire lives. The Eucharist "is an extremely personal encounter with the Lord and yet never simply an act of individual piety."[215] In the Eucharist we find a visible manifestation of the reality that we "are the Church of Christ, his members, his body. We are brothers and sisters in him. And in Christ, though many and diverse, we are one: *In Illo uno unum*."[216] The Eucharist opens us to justice and sharing, with a preferential concern for those who are

---

214. Saint Augustine, *Sermons*, 272: *In die Pentecostes ad infantes de sacramento*: *PL* 38, Paris 1865, col. 1247.

215. Benedict XVI, *Homily at the Mass of the Lord's Supper* (April 21, 2011): *AAS* 103 (2011), 321.

216. Leo XIV, *Address to the Roma Curia for the Exchange of Christmas Greetings* (December 22, 2025): *L'Osservatore Romano*, December 22, 2025, 6–7.

burdened by poverty or marginalization. And while new economic and technological networks can generate exclusion, isolation and dependencies, the Church—nourished by the Eucharist—is called to make visible a different paradigm, one that preserves human connections, gives a voice to the invisible and ensures that processes are aimed at respecting people's dignity.

The construction site of our time

236. The spirituality I wish to commend is that of the "wise architect" who, driven by hope for the Kingdom of God, is committed to building the world for the common good (cf. 1 Cor 3:10). As I mentioned at the beginning of this reflection,[217] the task of building in our time must place our relationship with God at its center. Our rule must be the acceptance of human limitations as a natural and positive reality, and should be characterized by shared responsibility and a language characterized by the Gospel. At the end of this reflection, the plan for a civilization of love can be seen more clearly, and the construction site appears to be already up and running, thanks especially to the many living stones solidly united to Christ the cornerstone (cf. 1 Pet 2:4–6). In this task, we are called to assume an active role, without taking refuge in spiritual sentimentality or retreating into our

217. Cf. above, nos. 11–14.

own little worlds. We must be faithful to the truth, invest in education, cultivate relationships and love justice and peace.

237. Let us remain faithful to the truth! Living amid incessant flows of information, opinions and images, we know how easy it can be to influence decisions and preferences through increasingly sophisticated algorithms.[218] In this context, it is imperative to cultivate hearts that love the truth, prefer what is right despite the most appealing content and pursue wisdom rather than immediate results. We must always keep before us the truth about God and humanity, just as Christ has revealed them to us. We must lay aside an individualistic and technical view of humanity, as if reality were mere matter to be shaped according to selfish interests, whether individual or collective.[219] Instead, let us cultivate what Pope Francis called a "situated anthropocentrism,"[220] which recognizes the human being as a creature embedded in a network of relationships with other living beings and with all of creation. Fidelity to the truth requires integrating the possibilities offered by technology within a framework marked by wisdom, which is capable of

---

218. Cf. Leo XIV, *Address to the Conference "The Dignity of Children and Adolescents in the Age of Artificial Intelligence"* (November 13, 2025): *L'Osservatore Romano*, November 13, 2025, 3.

219. Cf. Benedict XVI, Encyclical Letter *Caritas in Veritate* (June 29, 2009), 34: *AAS* 101 (2009), 668–670.

220. Francis, Apostolic Exhortation *Laudate Deum* (October 4, 2023), 67: *AAS* 115 (2023), 1059.

safeguarding both the dignity of each person and the future of our common home.

238. Let us invest in education, beginning with ourselves! We all need to learn how to engage with the digital world in a human way, as an integral part of our education in the faith and in a life lived according to the Gospel. Indeed, we must consider the digital world as a new continent to be evangelized, one that requires generous missionaries who are mature in the faith. In a particular way, we need adults to rediscover their vocation as artisans of education, prepared to work patiently each day, with the support of extensive and shared educational partnerships. Today, accompanying children and young people in using technology for developing responsible relationships, helping them to recognize the risks and choose what fosters inner freedom, is a concrete form of charity and will safeguard their dignity. Teaching new generations that technological evolution does not follow a predetermined path, but can be guided by personal and collective responsibility, constitutes one of the most valuable services to the common good.

239. Let us cultivate relationships! In an era that favors speed and fragmentation, the human person still yearns to receive care and recognition from attentive minds, kind words and hands capable of tenderness. The digital culture multiplies connections and offers new opportunities for interaction; yet, the human heart retains an irrevocable need for genuine closeness. I invite everyone to cherish places and times where physical presence remains crucial, such as shared

meals, Christian community gatherings, time spent with the lonely and serving the poor. These are signs of a humanity that continues to believe that every person's body is a dwelling place of God and a temple of the Holy Spirit. It is precisely this covenant between glory and fragility that becomes the criterion for evaluating the anthropological models offered by contemporary culture.

240. Let us love justice and peace! The same technologies that facilitate communication and access to resources can also support models that exploit the most vulnerable, create new forms of slavery and derive profit from conflict. Every technical or economic decision should include spiritual discernment and be an opportunity for assessing whether the advances in AI are promoting justice and participation or concentrating wealth and power in the hands of a select few. I would encourage a careful examination of the supply chains of digital production, the working conditions hidden behind our devices and the mechanisms that profit from manipulation and war. At the same time, practical ways of fostering fairness, participation and care for creation must be found. We proclaim a hope rooted in the One who came down from heaven to "create a new story here below." For this reason, those who believe are committed to ensuring that a greater justice will take the place of inequality, and that the industry of war will be replaced by the craft of peace.[221]

---

221. Cf. Leo XIV, *Angelus on the Solemnity of the Epiphany* (January 6, 2026): *L'Osservatore Romano*, January 7, 2026, 3.

241. As we look to the future, I would like to recall the image of Nehemiah whom we chose as our companion and guide at the outset. Nehemiah heard the cry of a devastated city, brought that pain to prayer, discerned before God, asked for help, received permission to return, organized the work, confronted internal and external resistance and rebuilt the walls of Jerusalem with the assistance of the people, brick by brick. In this era of digital transformation, I see in him a striking parable of our own vocation, which is not to be passive spectators of social and cultural fractures, nor mere commentators on what is crumbling, but men and women prepared to enter the construction sites of history—research laboratories, technology companies, schools, the media, institutions and local communities—in order to rebuild what has collapsed and protect what is threatened. Like Nehemiah, we too are called to unite listening and courage, prayer and responsibility, so that, even when a technocratic mentality or partisan interests seem to prevail, the human city may become a more fitting place to live.

242. The image of rebuilding Jerusalem evokes the New Testament promise of the holy city, which is given to us first and foremost as a gift. In the Book of Revelation, the new Jerusalem descends as a gift for all God's people, "prepared as a bride adorned for her husband" (Rev 21:2). The walls of Jerusalem are no longer defensive fortifications, but the precious adornments of the Bride of the Lamb. Its gates, which Nehemiah guarded so diligently, remain permanently open to

all nations. God's presence offers light and life to all. The city is a new Eden, with its living water offered to the thirsty, and its tree of life whose leaves "are for the healing of the nations" (Rev 22:2). As we await its fulfillment, this vision is set before us as an encouragement—a call to overcome our divisions and to work together—for this is the way of Jesus Christ, yesterday, today and forever.

The song of hope: the *Magnificat*

243. After having considered faith, which contemplates the Father's loving plan; love, which unites us in one ecclesial body; and hope, which sustains our actions in the world, the fourth pillar of this program for Christian life is prayer. Mary's song accompanies our commitment. Before Elizabeth who announces to her that she has become the mother of the Lord, Mary bursts into a hymn of praise and joy. Her soul magnifies the Lord, and her spirit rejoices in God her Savior, for he chose a young, poor and humble girl for his plan of salvation. Mary suddenly sees all of history through the lens of this revelation. Nothing has changed around her; the socio-political situation of her time remains the same. The Romans continue to control her land, and her people are still subjugated and humiliated. Yet, everything has changed within her, and this allows her to see what is invisible. God has *already* shown the strength of his arm; he has *already*

scattered the proud, cast down the mighty, lifted up the lowly, filled the hungry with good things and sent the rich away empty-handed. He has *already* helped Israel, his servant. God "takes the part of the lowly. His plan is one that is often hidden beneath the opaque context of human events that see 'the proud, the mighty and the rich' triumph. Yet his secret strength is destined in the end to be revealed."[222]

244. The Blessed Virgin Mary not only teaches us to recognize God's invisible work, but also directs our gaze to "the points at which humanity is broken and the world becomes distorted: the contrast between the humble and the powerful, the poor and the rich, the satiated and the hungry," teaching us "to look at the world from a lower position: through the eyes of those who suffer rather than the mighty; to view history through the eyes of the little ones, rather than through the perspective of the powerful; to interpret the events of history from the viewpoint of the widow, the orphan, the stranger, the wounded child, the exile and the fugitive."[223] The Blessed Virgin thus becomes "poet and prophetess of Redemption," because on her lips is proclaimed "the strongest and most innovative hymn ever articulated, the *Magnificat*; it

---

222. Benedict XVI, *General Audience* (February 15, 2006): *L'Osservatore Romano*, February 16, 2006, 4.

223. Leo XIV, *Meditation on the Occasion of the Prayer Vigil and Rosary for Peace* (October 11, 2025): *L'Osservatore Romano*, October 13, 2025, 2.

is she who reveals the transformative vision of the Christian economy, the historical and social result that still draws its origin and strength from Christianity."[224]

245. With the same faith as Mary, let us become "weavers of hope" in our world, sharing who we are and what we have, so that the presence of Jesus may grow among us and his Kingdom take shape. In the humble fidelity of daily life, even the era of AI can become a time in which the Holy Spirit brings about the civilization of love in our lives. Indeed, the Lord continues to make all things new and offers every era the possibility of becoming part of salvation history in the light of the Incarnation. I entrust our desire to the Mother of Christ, to the Woman of the *Magnificat*, that she may guide our steps through this time of change and preserve in each of us true faith in the Gospel, so that we may bear witness to the grandeur of humanity, in which God has made his dwelling.

Given in Rome, at Saint Peter's, on May 15, in the year 2026, the second of my Pontificate.

LEO PP. XIV

---

224. Saint Paul VI, *Homily at the Marian Shrine of Our Lady of Bonaria* (April 24, 1970): *AAS* 62 (1970), 301.

Thank You.

Your purchase of this book and engagement with our other projects supports us in the work we do as Daughters of St. Paul. This book is the fruit of our consecrated life, prayer, and mission of communicating God's love.

We hold you and all your intentions in our prayers. We invite you to connect with us or send us prayer intentions at pauline.org.

Pauline
BOOKS & MEDIA